The Advertising of

Massey-Harris, Ferguson and Massey Ferguson

The Advertising of Massey-Harris, Ferguson and Massey Ferguson

John Farnworth

FARMING PRESS

First published 1999

ISBN 0 85236 528 4

A catalogue record for this book is available from the British Library

Published by Farming Press
Miller Freeman UK
2 Wharfedate Road
Ipswich IP1 4LG
UK

Distributed in North America by
Diamond Farm Enterprises
Box 537, Bailey Settlement Road
Alexandria Bay, NY 13607, USA

Designed by Hannah Berridge
Edited by Hal Norman

Printed in Hong Kong through World Print Ltd

Contents

Acknowledgements

I am most grateful to AGCO, the holders of the Massey Ferguson brand name, for allowing me to reproduce this wide selection of advertising material.

John Burge, Rory Day, Richard Ghent, David Lory, Ken Reichert, Malcolm Robinson and Jim Russell have all generously helped me to collect together specific pieces of literature which I needed to make a complete representative sample of the whole period. Thanks also to Beverly Hughes who kindly arranged for the copying of some material in Canada and to Lynn Campbell who researched some of the early Massey, Harris, Massey-Harris and Ferguson literature from the archives of the Ontario Farm Museum. I also wish to record my gratitude to the many vintage machinery enthusiasts with whom I have discussed the idea and who encouraged me to undertake the compilation.

Finally, thanks must go to my son Trystan who one morning woke up with the idea for this book, and to my wife Moira for help with editing and typing up all the captions.

Foreword

In compiling this advertising history of Massey Ferguson and its predecessor companies, John Farnworth has made a unique historical and global record of the company's product range through over 150 years. He has achieved this by gathering together a highly illustrative and colourful collection of advertising material from early Massey days through to the present day. The selection he has chosen has been thoroughly researched to ensure a representative coverage of our advertising history in terms of both the advertising styles which have been adopted over the years and the range of principal company products.

The book is in many ways an illustrated chronicle of farm mechanization through the eyes of one company. In the range of advertising material presented we see a progression from highly detailed and colourful hand artwork through to camera, and then on to computer-assisted camera work.

Many trends are seen in this progress of mechanization, such as a shift from horse-drawn to tractor-drawn and then engine-powered machinery, and a shift from the small implements of the horse era to today's gigantic tractors, implements and combine harvesters. We can also clearly see the demise of the small family farm characterized in its idyllic rural settings as witnessed by our loss of emphasis on this since about the 1960s. We also see a trend in the advertisements from the depiction of man plus horse plus machine working in harmony to man plus tractor working in a new era of tractor-oriented happiness. However, since sometime in the 1960s advertising has moved away from the tractor plus man theme towards an emphasis on the sheer power and size of modern farm machinery in which man is still represented in the advertisements but is not so strongly emphasized. Indeed, as the agricultural industry relentlessly proceeds down the big business highway, it is perhaps a source of regret that emphasis on the rural idyll and personality of the machine operators (and even the horses) has been lost from the advertising messages.

John has brought the advertising material together in a very organized manner which allows us to cruise through this glorious century and a half of progress in farm mechanization. He has included our main and some lesser known product lines. Each advertisement is accompanied by an informative caption which adds immensely to the general reference value of this book. Yesterday's artists and photographers would be well satisfied to see their efforts paraded in such manner and indeed today's advertisers may well sit back and take stock of their methodology when they peruse this book!

The book will be of interest to those interested in agricultural history and farm mechanization, vintage tractor and machinery enthusiasts and indeed anyone with an interest in the countryside. It will also appeal to artists and advertisers in these fields.

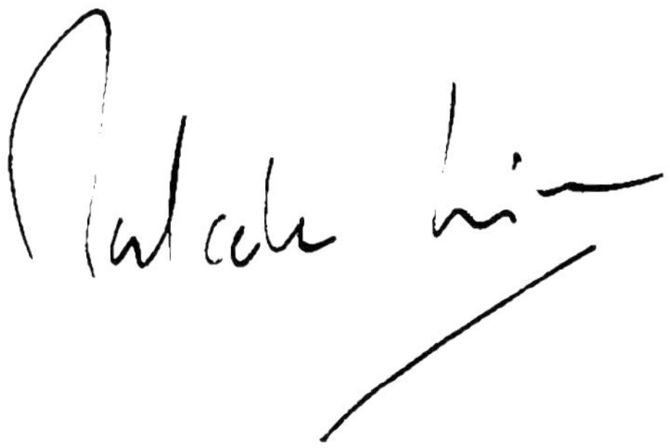

Malcolm Lines, Vice President, Marketing
AGCO European, Coventry, UK

Introduction

It is now over a century and a half since Daniel Massey founded the company that became known as Massey Ferguson in 1957. This company continued until late into the 1990s when Massey Ferguson became a registered brand name of AGCO, thus ensuring that the manufacture of products under the Massey Ferguson banner, and the name and the legend live on.

The last 150 years have witnessed vast changes in the agricultural industry worldwide, most particularly in the western world. Much of this development has been prompted by the evolution of ever higher output farm machinery. This in turn has seen significant changes in the social structure of agriculture and sadly, perhaps, the demise of the romantic appeal of farming and the old community nature of the industry. The period has also seen tremendous development in marketing strategies as print advertisements have become increasingly sophisticated, both in their psychology and technology.

Massey Ferguson is one of the world's oldest international agricultural machinery companies. Its development has been, in effect, not only a history of an international machinery company, but also a history of agricultural machinery development and marketing. Printed advertising material has been used to support the marketing effort since very early days.

I have endeavoured to draw together a selected range of advertising material from early company days to the present in order to reflect the changing nature of company products, the changing focus of advertising messages, changing printing and associated artistic techniques and the changing social nature of agriculture. I have also sought to record, through advertisements, at least some of the companies taken over during the course of the corporate development of Massey. Perhaps the most significant amalgamation of companies in this long history is that of Massey-Harris and Ferguson in 1953.

Simple observation of the changing nature of the company's advertising, and the information which it delivers, has increasingly come to provide me with a fascinating insight into the world of agricultural machinery. In short, it has become an inseparable and complementary activity to the collection of, and general interest in, vintage agricultural machinery.

Thousands of different pamphlets, brochures and posters have been produced over the years. There is no complete archive in existence and many earlier ones have undoubtedly been lost for all time. Fortunately, their collection is becoming a popular activity and this means that their preservation is ensured. For many, literature collection has become complementary to the collection of machinery. The vast amount of material which has been produced over time means that a book like this can only provide a representative selection of relevant material. The choice is mostly mine, but with a significant contribution from those who have loaned material and to whom I simply asked to 'find me some of the more interesting and different specimens' to add to the overall review which I have tried to create.

The selection of advertising material which comprises this book is mostly, but not exclusively, front cover material of advertising leaflets, brochures and catalogues, and a few posters. Some of the photos show blemishes of one form or another – sadly, an inevitability with older material. A few are slightly blurred or dark but this is how they originally appeared. Wherever possible I have given the date of the literature. For convenience I have divided the book into five periods – Massey, Harris, Massey-Harris, Ferguson and Massey Ferguson. Each period is sub-divided in order to reflect either development eras or product ranges. The dates are shown on the section heading.

Section I

Massey 1847–1891

In 1847 Daniel Massey, a descendant of 17th century English immigrants to Canada, established a small business in Ontario which became, as Massey-Harris, the largest agricultural equipment manufacturer in the British Empire. It subsequently became Massey Ferguson, one of the world's foremost agricultural equipment firms surviving until the late twentieth century. In the 1990s Massey Ferguson was taken over by AGCO who continue to produce and market a large range of Massey Ferguson tractors and equipment. However, the name 'Massey' really does live on as it is the commonly used term of reference for many of the company's products.

In 1870 the Massey Manufacturing Co. was formed, under the management of Daniel's son and successor, giving the business its first corporate identity.

Records show that the first Massey catalogue was issued in 1857, a year after Daniel's death. Printed advertising material was clearly a success as Massey established its own printing plant in 1885.

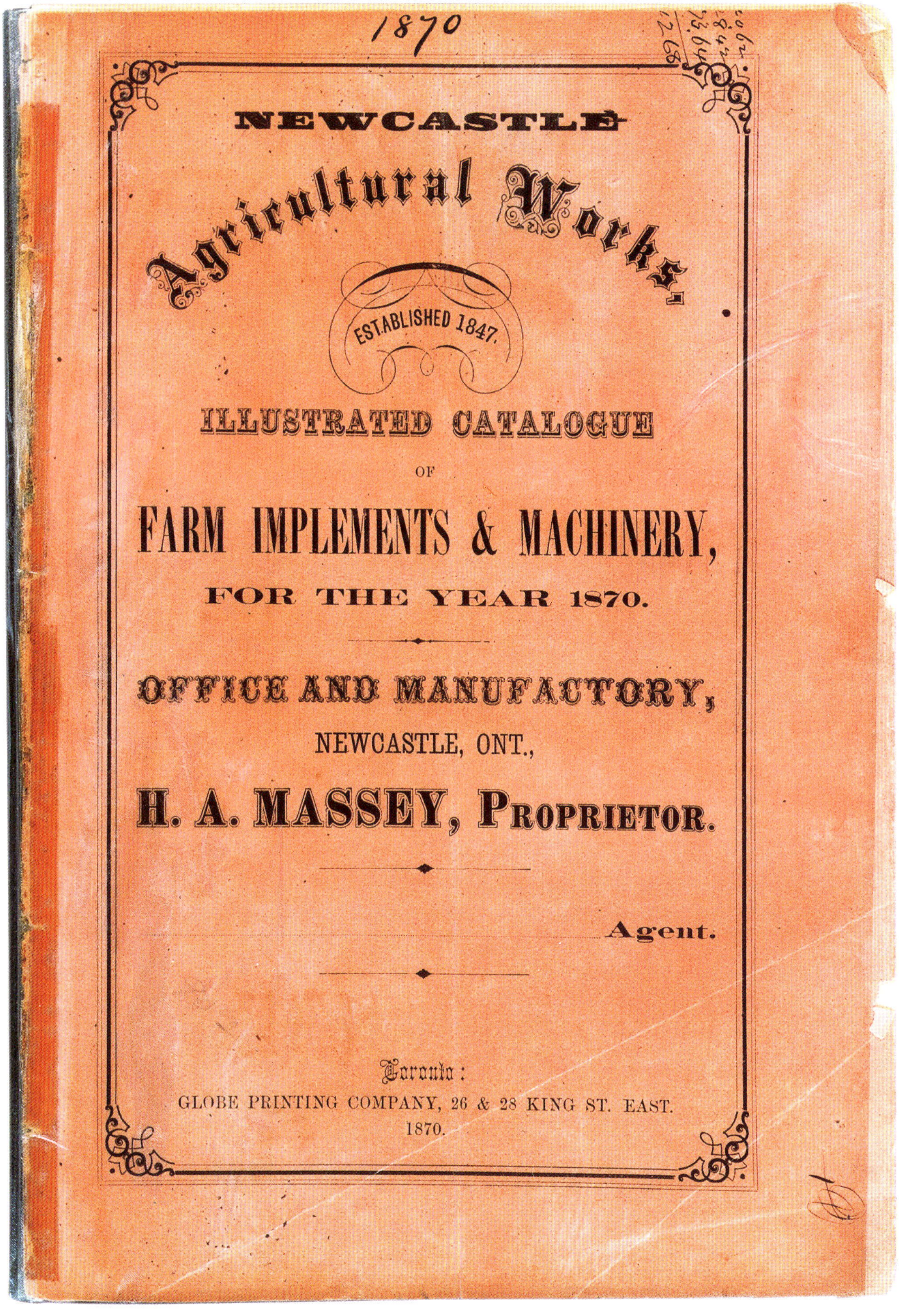

Daniel Massey acquired his first major works in 1848 at Newcastle, Ontario, in only the second year of the company's history. This 1870 catalogue shows that the company had transferred to his son, Hart Massey. Nine years later Hart founded the massive Toronto factory. Also in 1870, the company changed its name to the Massey Manufacturing Co.

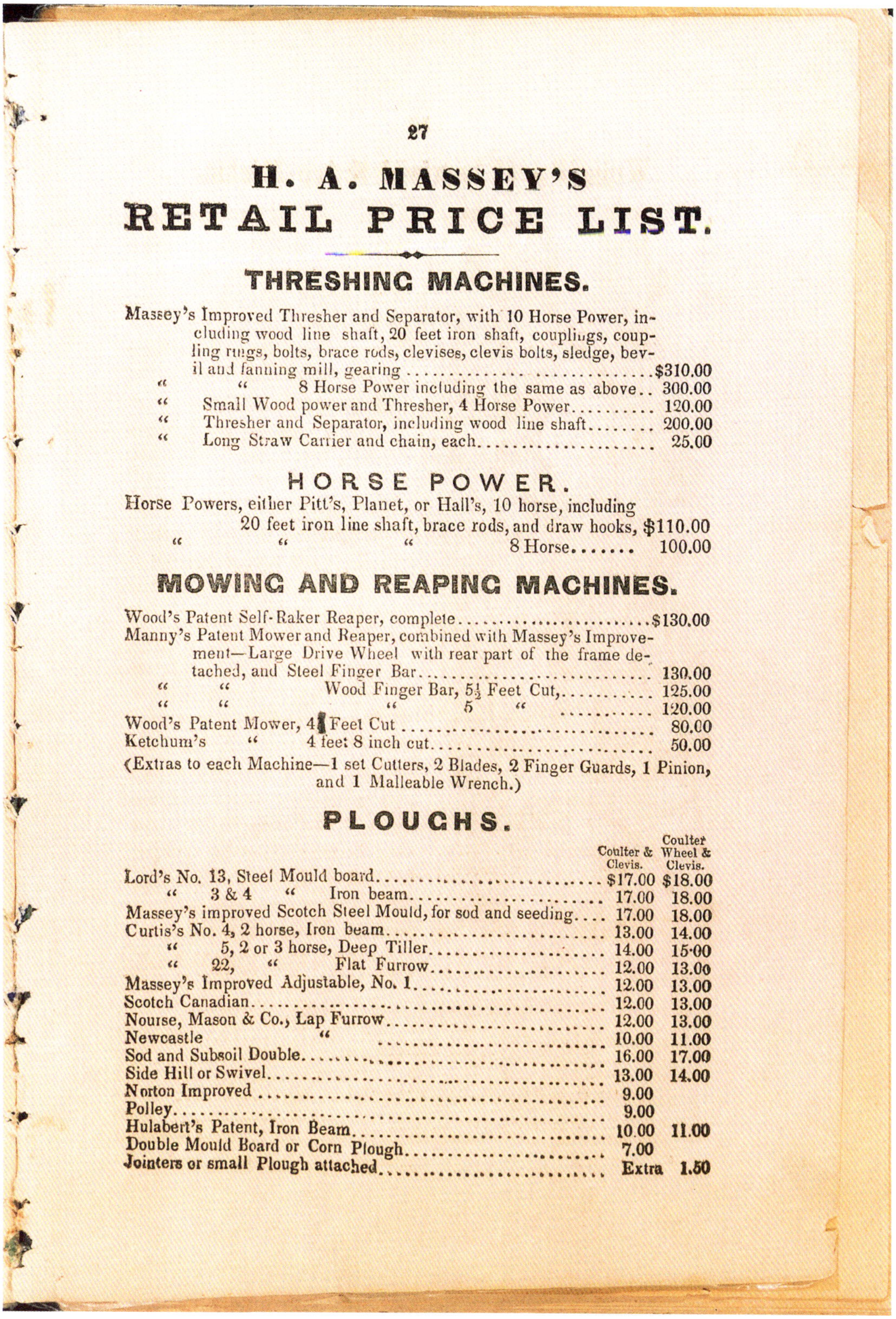

27

H. A. MASSEY'S RETAIL PRICE LIST.

THRESHING MACHINES.

Massey's Improved Thresher and Separator, with 10 Horse Power, including wood line shaft, 20 feet iron shaft, couplings, coupling rings, bolts, brace rods, clevises, clevis bolts, sledge, bevil and fanning mill, gearing $310.00
" " 8 Horse Power including the same as above .. 300.00
" Small Wood power and Thresher, 4 Horse Power 120.00
" Thresher and Separator, including wood line shaft 200.00
" Long Straw Carrier and chain, each 25.00

HORSE POWER.

Horse Powers, either Pitt's, Planet, or Hall's, 10 horse, including 20 feet iron line shaft, brace rods, and draw hooks, $110.00
" " " 8 Horse 100.00

MOWING AND REAPING MACHINES.

Wood's Patent Self-Raker Reaper, complete $130.00
Manny's Patent Mower and Reaper, combined with Massey's Improvement—Large Drive Wheel with rear part of the frame detached, and Steel Finger Bar 130.00
" " Wood Finger Bar, 5½ Feet Cut, ... 125.00
" " " 5 " 120.00
Wood's Patent Mower, 4[illegible] Feet Cut 80.00
Ketchum's " 4 feet 8 inch cut 50.00

(Extras to each Machine—1 set Cutters, 2 Blades, 2 Finger Guards, 1 Pinion, and 1 Malleable Wrench.)

PLOUGHS.

	Coulter & Clevis.	Coulter Wheel & Clevis.
Lord's No. 13, Steel Mould board	$17.00	$18.00
" 3 & 4 " Iron beam	17.00	18.00
Massey's improved Scotch Steel Mould, for sod and seeding	17.00	18.00
Curtis's No. 4, 2 horse, Iron beam	13.00	14.00
" 5, 2 or 3 horse, Deep Tiller	14.00	15.00
" 22, " Flat Furrow	12.00	13.00
Massey's Improved Adjustable, No. 1	12.00	13.00
Scotch Canadian	12.00	13.00
Nourse, Mason & Co., Lap Furrow	12.00	13.00
Newcastle "	10.00	11.00
Sod and Subsoil Double	16.00	17.00
Side Hill or Swivel	13.00	14.00
Norton Improved	9.00	
Polley	9.00	
Hulabert's Patent, Iron Beam	10.00	11.00
Double Mould Board or Corn Plough	7.00	
Jointers or small Plough attached	Extra	1.50

The first Massey catalogue was issued in 1857. This page from an 1862 catalogue is interesting because it lists three of Massey's first and most important 'acquired products' which launched the company into the harvesting machinery sector. They are the Ketchum mower (1851), the Manny reaper (1855) and Wood's reaper (1861). As can be seen from the list, there were obviously other 'acquired' products about which little is known.

THE MASSEY MANUFACTURING CO. 21

IMPROVED THRESHER AND SEPARATOR.

Our constant aim shall be, in building this machine, to spare no pains or expense in making it meet the wants of the public. We do this the more cheerfully because of the universal favour which it has received in every neighbourhood where it has been introduced, availing ourselves, too, of the friendly suggestions of experienced threshers that we occasionally receive, as far as practicable.

Those acquainted with the machine will observe some changes on those to be sent out in future, that will add considerably to the convenience of threshers. From year to year the demand for Threshing Machines has steadily increased, and for several seasons we have been unable to fill all our orders. Now, with increased facilities and additional improved machinery, that will enable us not only to produce a greater number than heretofore, but to add to the already acquired reputation for superior workmanship and finish, no effort will be spared to maintain it before the community as one of the best threshers in use.

We think that the following reasons, among others that might be named, should influence every one desiring a Thresher and Separator to give our's the preference:—

The frame is made heavier, and more thoroughly braced than usual.

We use steel shafting in many parts, and have lengthened some of the bearings.

We make a wider Wheat Belt and Sieves than ordinarily used.

We furnish the best Patent Stretched Leather Belting.

We know no machine more durable, and less liable to get out of repair.

It will thresh and clean, fit for market, as much grain, in as quick time, and do it as well as any other known:—in good grain-growing sections 100 bushels per hour being considered its average capacity.

There is no other machine known that will do the same amount of work with less power, when kept in proper condition.

Daniel Massey is reputed to have brought the first threshing machine into Canada (from America) before he started manufacturing agricultural equipment in 1847. He used the thresher for contracting purposes. This 1870 Massey Manufacturing Co. catalogue shows an early Massey-produced thresher (badged H. A. Massey – the name of the company changed in this year).

In 1875, Massey purchased the patents for Ontario, Quebec and Manitoba to make the Sharp's Rake. Manufactured in Newcastle and later in Toronto, it was an important item in the company's line of hay tools and was exported extensively.

The Massey Manufacturing Co. acquired the rights for the Sharp's self dumping rake in about 1875. The machine had one of the longest production runs of any piece of Massey equipment. The advertisement appeared in 1881.

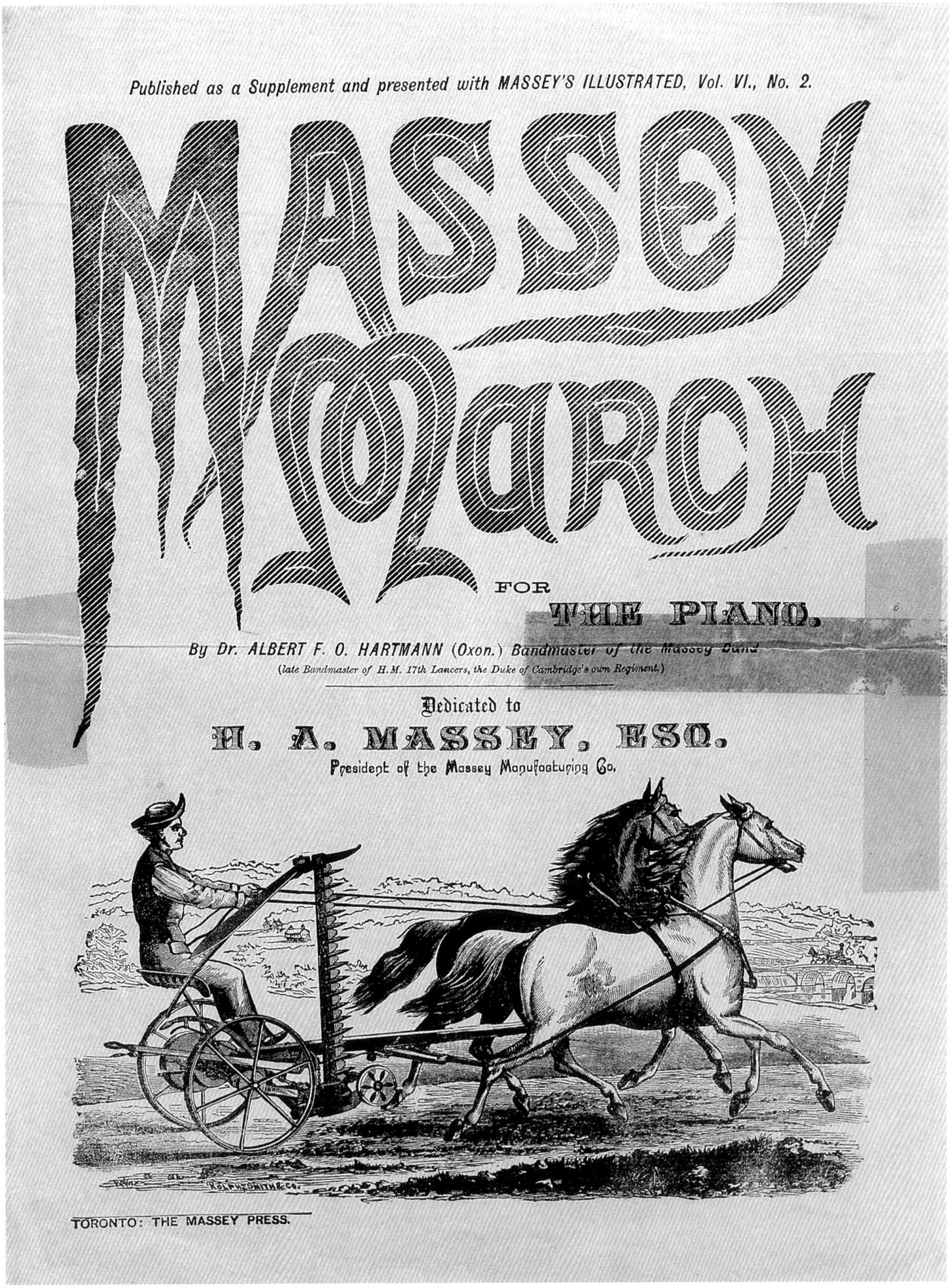

A Massey Band and a Massey Orchestra were formed soon after the start of the Toronto factory. The Toronto factory was used for employees' social events. This supplement to the *Massey Illustrated* magazine, produced by the Massey Press in the 1880s, shows that a Massey March had been composed and dedicated to Hart Massey.

The Toronto Reaper and Mower Company, makers of the Toronto Reaper, was acquired by Massey in 1881.

The purchase by Massey of the Toronto Reaper and Mower Co. was one of Massey's most significant acquisitions, giving Massey a very successful binder, as well as a mower range. 1881.

Building on success – the Toronto Mower production increased year on year. An interesting case of text and picture being presented in two different orientations! 1882.

The Massey Low Down Binder was one of the highly successful machines which came with Massey's acquisition of the Toronto Reaper and Mower Co.

TORONTO.] SECOND EDITION. [FEB. 15, 1887.

OUR FLAG GOES UP
FIFTEEN FEET HIGHER.

WHITE & GOLD MACHINES EVERYWHERE
AWAY AHEAD.

Great Strategic Movements of the Enemy

THEY ADOPT THE "GIVE-AWAY" PLAN.

But the attempted seduction of the Public by false arguments and general palaver doesn't work.

The time of "SAW-LOG BINDERS" and "POT-METAL MACHINES" is gone by.

Canadian Farmers must and will have
MODERN MACHINES.

The whole superstructure of the Competition rests on gas-bag expostulations.

THEY DON'T STAND ANY SHOW AT ALL.

A long pull, a strong pull, and a pull altogether, and you'll [illegible]

The *White and Gold* magazine was a Massey newsletter produced for their agents. This issue is dated 1887, a time of intense competition in the binder market, with international exports going as far afield as Europe, Australia and New Zealand.

An idyllic family farm setting for 1888.

One of the last Massey Manufacturing Co. brochures to be produced prior to the merger with Harris. The sickle and scythe appear to have been abandoned once and for all, as fleets of binders take over the harvest of large cereal growing areas.

THE WORLD'S FAMOUS MASSEY-TORONTO LIGHT BINDER.

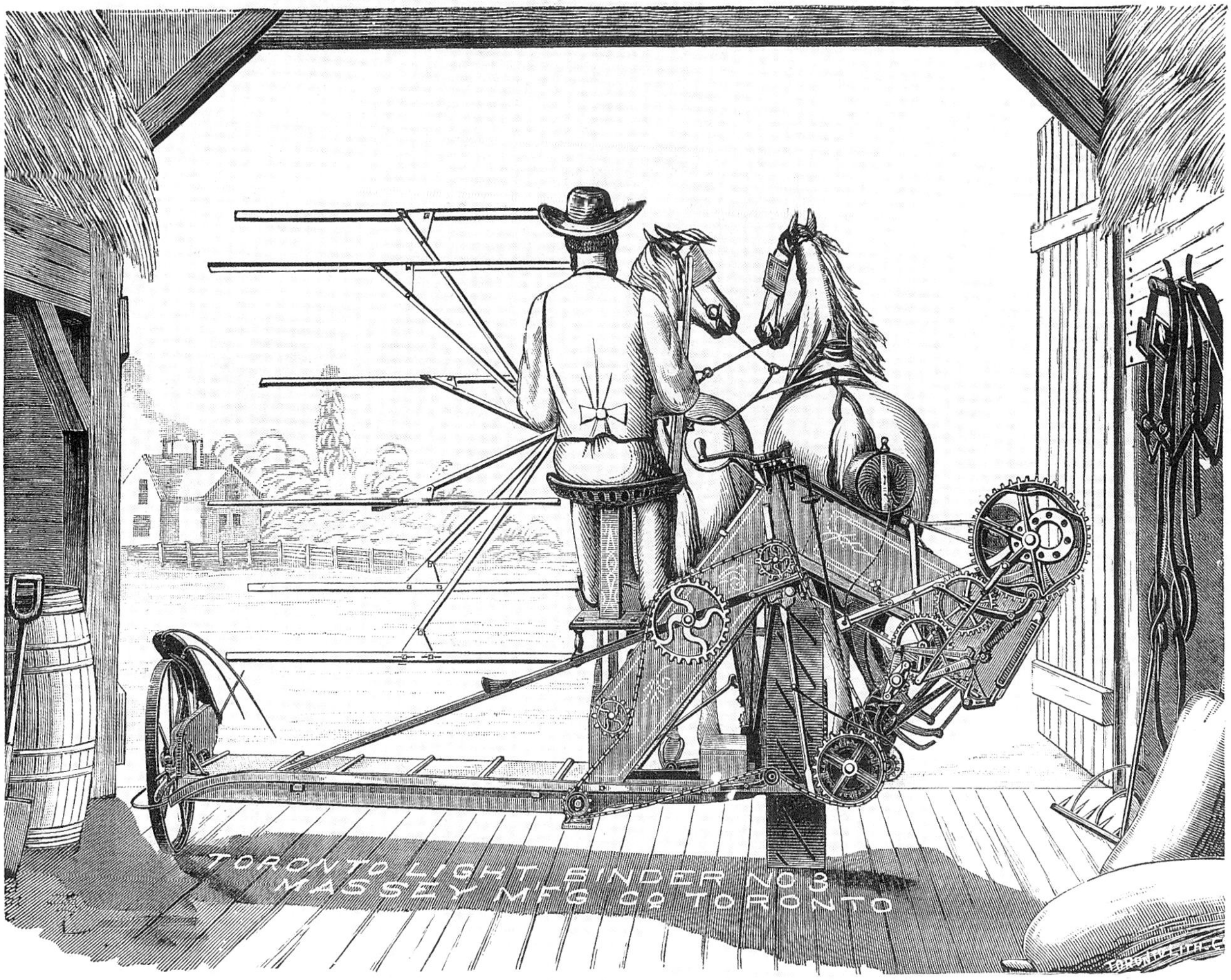

Going to Work with the TORONTO LIGHT BINDER No. 3A.

DOOR ONLY ELEVEN FEET WIDE.

nearly all the gearing of the ma-
iew. The simplicity of this cele-

Our No. 4A Binder is of exactl
as the 3A, except that it is wide

The Toronto Light Binder was one of Massey's best selling binders. It was produced up to and after the merger with Harris. 1891.

Section II

Harris 1857–1891

Ten years after Daniel Massey started his business, Alanson Harris, who was also a descendant of British immigrants to Canada, started a similar business which was to parallel Massey's. Harris's company was also based in Ontario.

Harris's business was so successful that he became the main competitor to Massey even though Harris was smaller in the binder business. Both companies developed their export trade and showed or demonstrated their equipment at major exhibitions. The two companies became locked in what has become popularly known as 'the battle of the binders', such was the demand for these machines at the time, and the competition to supply them.

Alanson Harris established his first major manufacturing plant at Brantford in 1872 and doubled its size in 1877. This 1880 catalogue shows Harris as a specialist in harvesting equipment.

Harris started selling Osborne wire tying binders in 1878 and by 1880 was advertising this Osborne twine binder.

Reapers and binders were sold alongside each other for many years. Harris's first reaper – the Kirby hand rake reaper was acquired from the Osborne Co. in 1863. Reel rake reapers like this one were introduced in 1870. 1883.

By the time this advertisement appeared in 1883 there was intense competition between Harris and Massey in the binder market. This competition prompted some very colourful advertising.

Design of the Brantford twine tie binder started in 1881. Harrisís 1886 *Journal for Farmers* shows a Brantford twine tie binder.

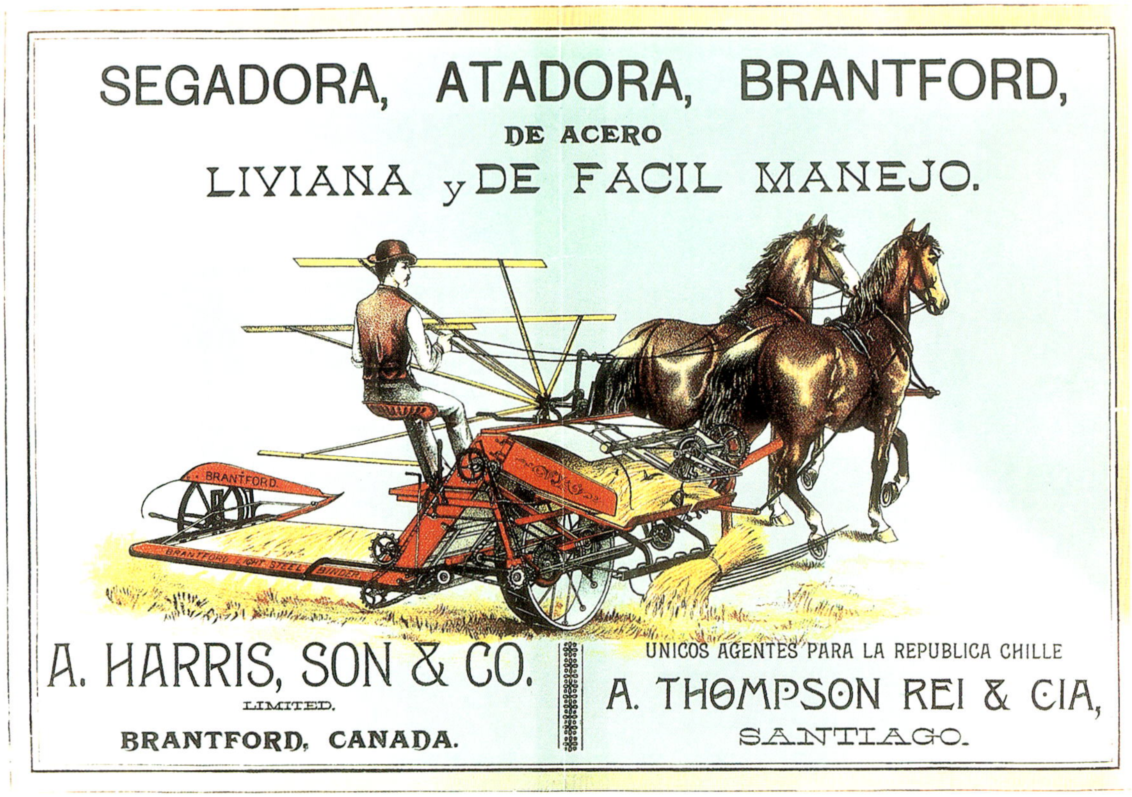

Just as Massey had developed international markets, so did Harris. This example of a Harris agent in Chile advertising a Brantford binder is dated 1888.

Brantford steel binders scored highly in international competition with Massey binders. So successful were they that it would seem they created some free time for country ladies! 1889.

: : THE BRANTFORD LIGHT STEEL BINDER No. 2 : : :

: WITH NEW BRANTFORD OR APPLEBY KNOTTER, AS PREFERRED :

By 1891, the year of the merger with Massey, optional knotters were available on the Toronto Light Binder.

Ease of transport and storage were key selling features of the Brantford binder.

Harris started his production of machines at Beamsville but later moved to Brantford – the town which gave its name to Harris's highly successful line of mowers, binders and reapers. 1891.

Production of reapers carried on in parallel with binders for many years. This Brantford reaper was still being sold in the Massey-Harris era. 1891.

Possibly the last catalogue of the Harris era. It is dated 1891, the year of the merger with Massey.

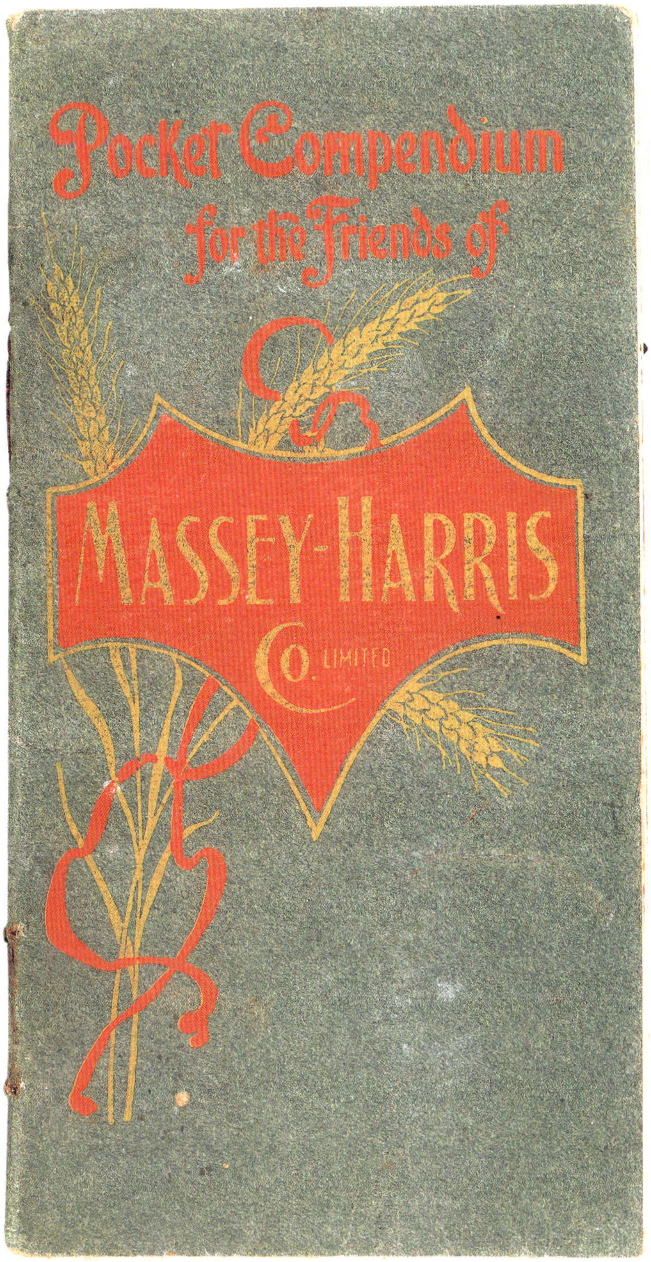

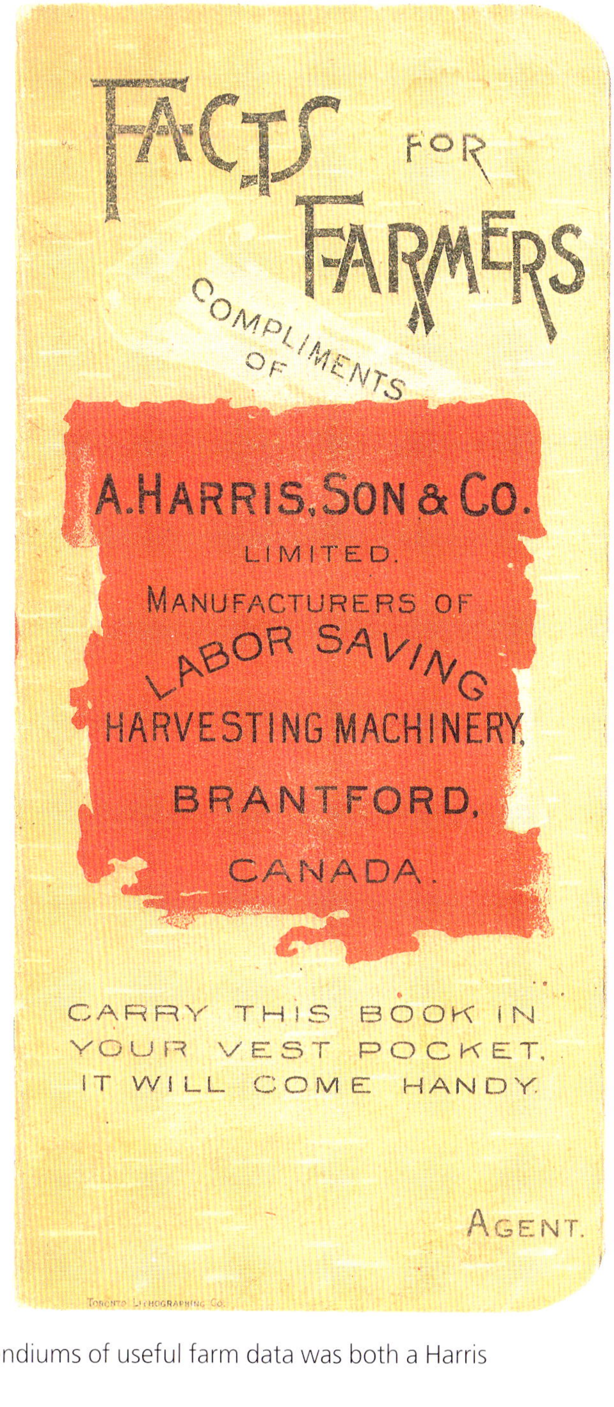

Advertising via small (2.5 x 5.5in. approx) compendiums of useful farm data was both a Harris and, later, a Massey-Harris strategy.

Section III

Massey-Harris
1891-1953

On 22 July 1891 Massey and Harris merged to form the Massey-Harris Company Ltd.

Prior to the merger the two companies' success had much to do with their recognition of the good products being made by other manufacturers which they then acquired the rights to market or produce. After the merger a spate of similar acquisitions followed. The products acquired considerably broadened the product range which Massey-Harris could offer. Throughout the lifetime of Massey-Harris this pattern continued, culminating in the merger with Ferguson in 1953. The Ferguson merger and the purchase of the Wallis tractor company in 1928 gave Massey-Harris important tractor lines.

Between 1891 and 1953 Massey-Harris became a major international company which not only marketed worldwide, but established significant production or assembly facilities in countries outside Canada.

The pivotal role that Massey-Harris played in the development of mass produced self propelled combine harvesters will never be forgotten.

Massey Illustrated magazine started in the Massey era. This back cover of the August 1891 edition announced the Massey and Harris merger.

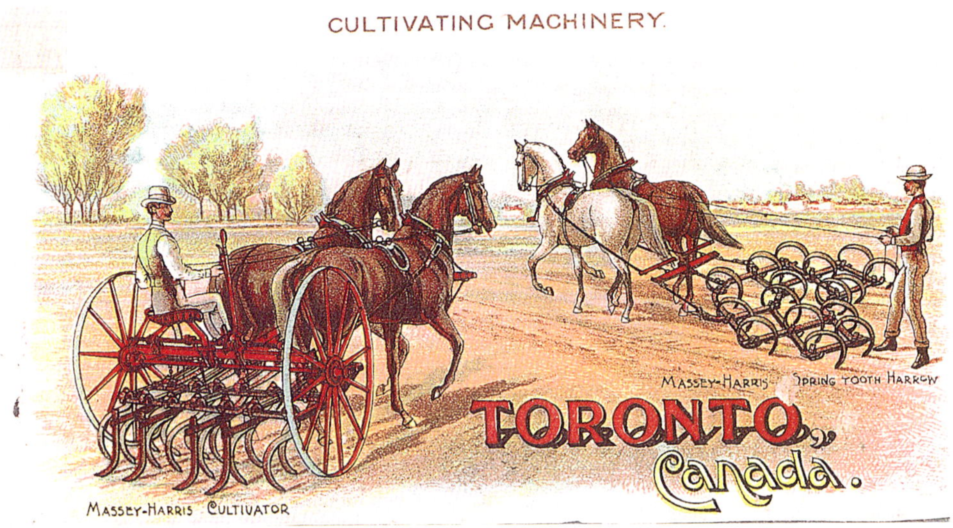

It is thought that these fine works of art were given out at Expos.

This machine has made its record and still leads. It is the best "all round" Reaper on the market to-day. Light and strong in design and construction. Wrought Iron Frame. Steel Shafting. Brass Bushing. Hoisting and Tilting Levers. All that can be desired. Works well in either standing or lodged grain.

La Moissonneuse "Massey."

Cinq Pieds de Coupe. Pour Deux Chevaux.

Four years after the Massey and Harris merger, Massey reapers were still being sold to France. 1895.

THE PATTERSON LIGHT STEEL BINDER
WITH IMPROVED KNOTTER

WOODSTOCK, ONTARIO

—5—

THE PATTERSON & BRO. CO.

The Lightest Weight and the Lightest Draft of any machine of the kind in the Market.
Perfectly Balanced. Superior in Manufacture, Finish, Durability and Efficiency.
Transporting Trucks and Sheaf Carrier furnished when required.

Patterson and Co. of Woodstock (who had only recently taken over J. O. Wisner, Son and Co. of Brantford) were the first company taken over by Massey-Harris shortly after their merger in 1891. Pattersons brought seeding and haying equipment to the M-H portfolio, whilst Wisner added seeding and tillage equipment. The advertisement can only be dated post-1889.

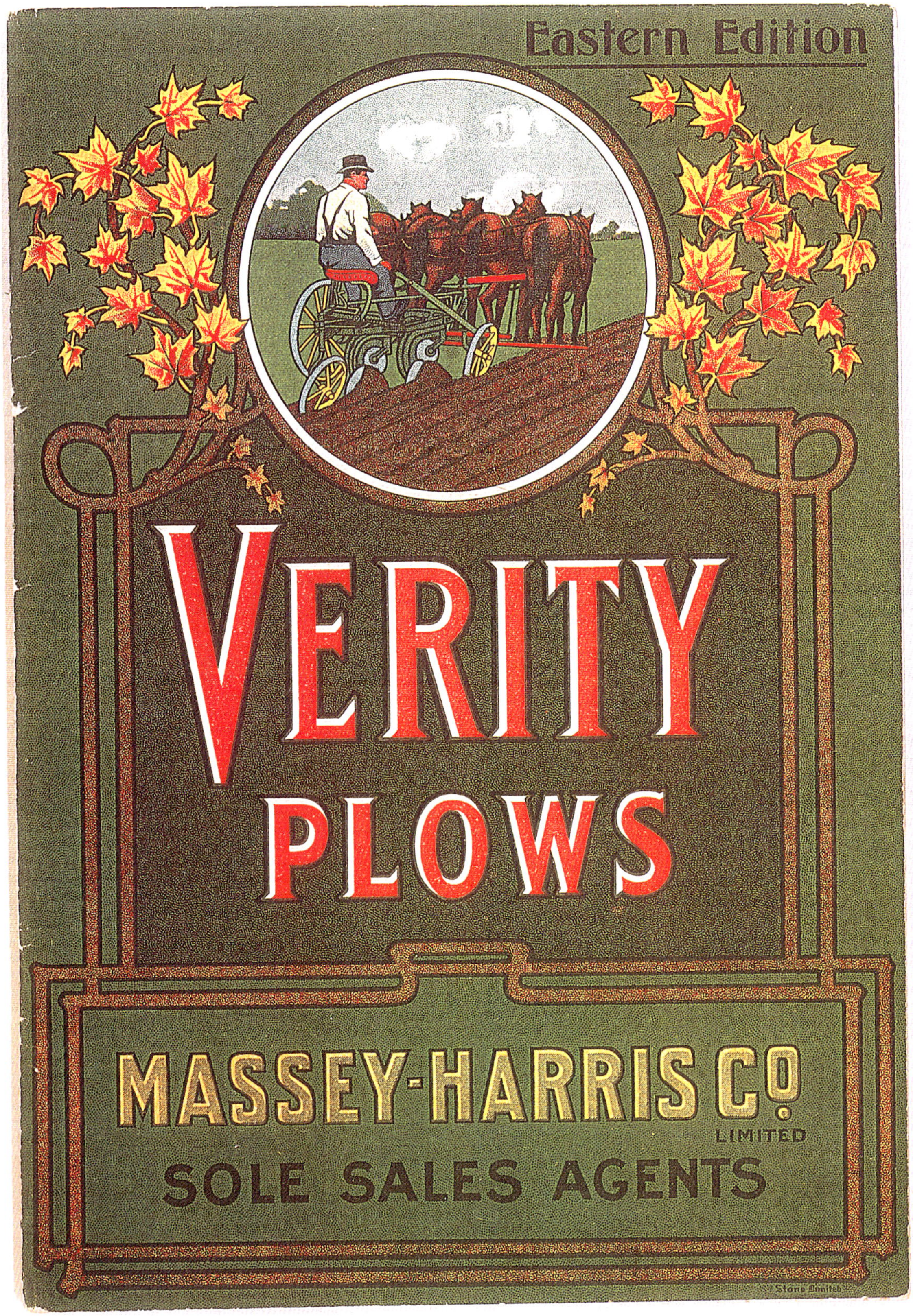

In 1892 M-H acquired the Verity Plow Co. of Exeter, Ontario. This gave M-H an extensive range of disc and mouldboard ploughs for anything from a horse to a massive steam traction engine. This late 19th century Verity M-H catalogue shows a horse-drawn two furrow Sulky plough.

This 12 disc plough is an example of one made for the large Prairie-type tractors.

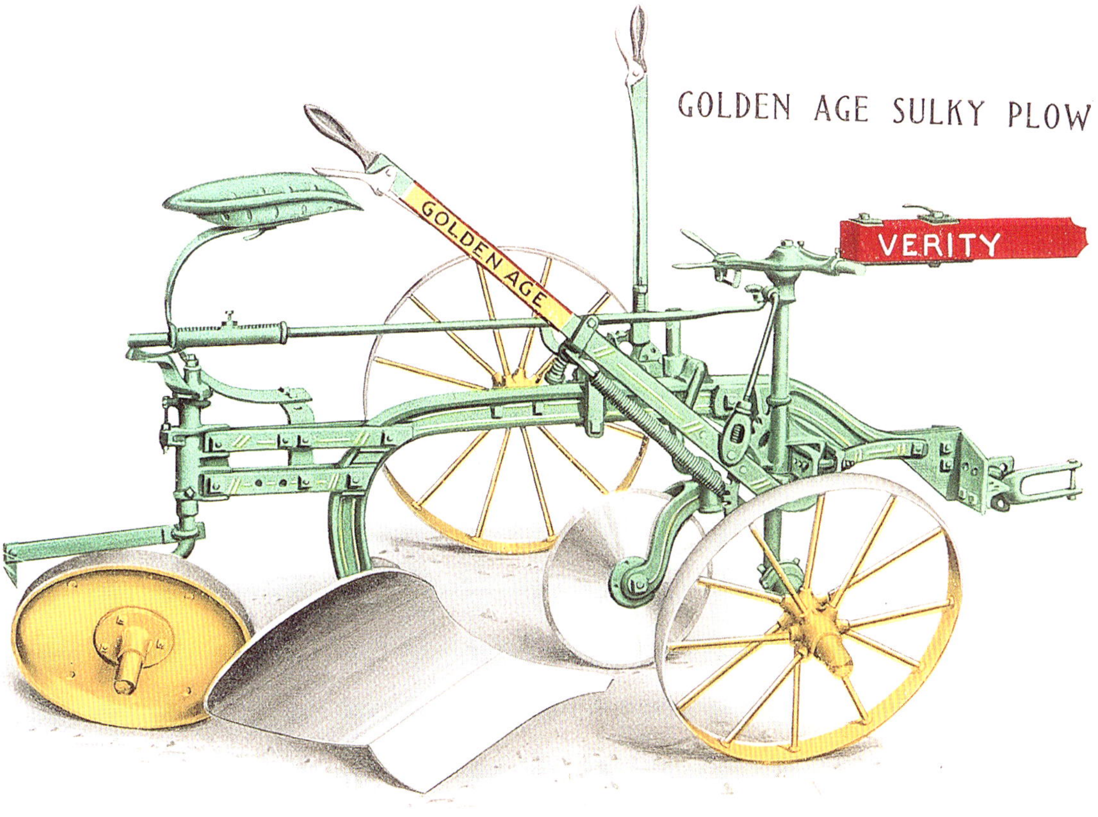

Produced for a single light horse, the Verity ride on Golden Age Sulky plough.

Verity also made a range of garden scufflers to complement the plough range. The Verity brand name was retained after their acquisition by M-H. The advertisement comes from 1897.

This M-H seeder with grass box is probably pre-1900.

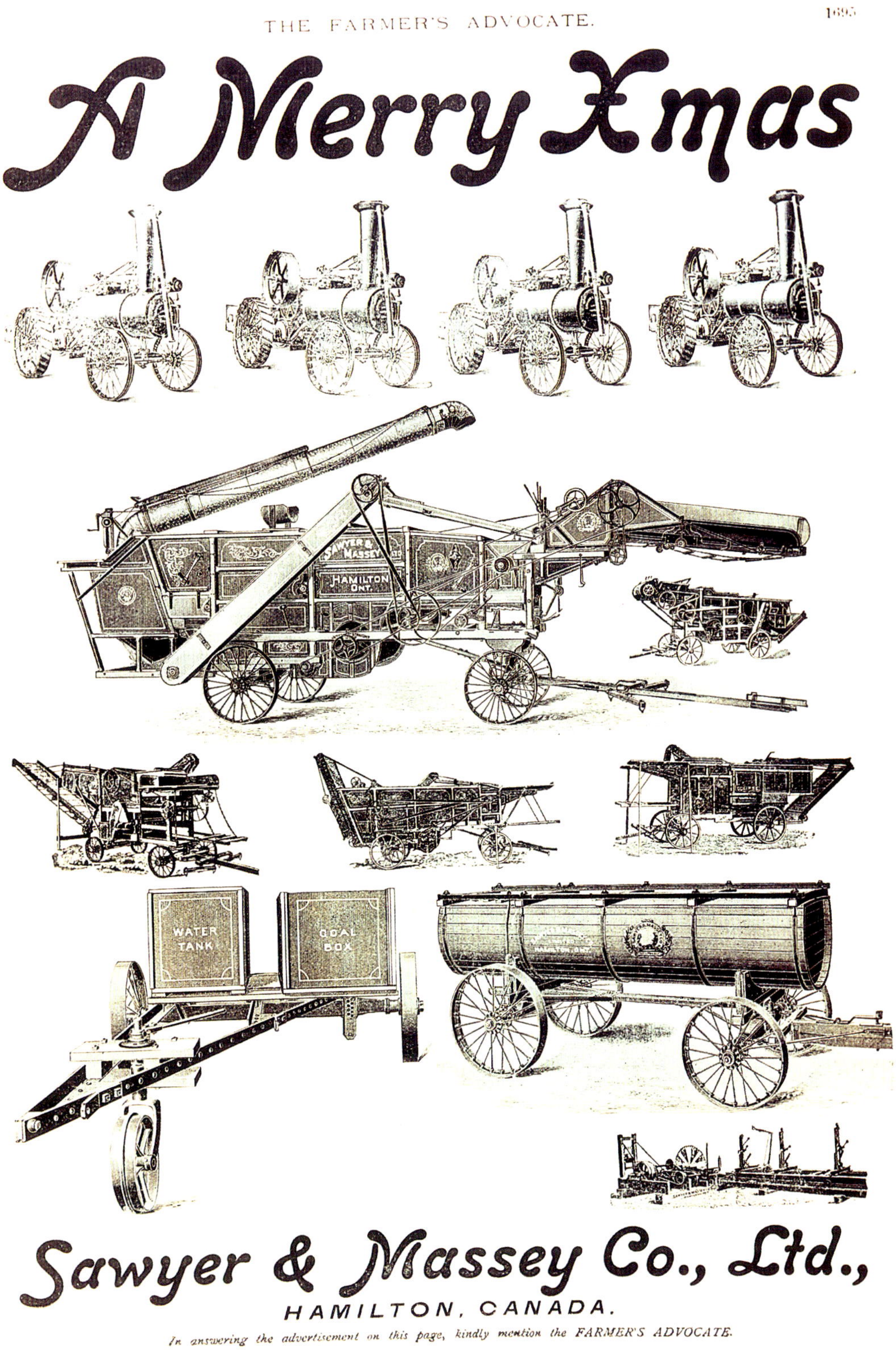

Some of the Massey family took 40 per cent of the shares in the L. D. Sawyer Co. of Hamilton in 1892 after years of informal relationships with Sawyer. However, no corporate link was established with the newly formed Massey-Harris. Sawyer specialized in steam tractors and engines, threshers and road making equipment, and later gasoline tractors. The Massey family withdrew from the company in 1910. 15 December 1904.

INVEST YOUR MONEY NOW IN

SAWYER MASSEY
HEAD OFFICE AND FACTORY
HAMILTON
ONTARIO
CAN.

MONEY-MAKING MACHINERY.

Sawyer-Massey Engines & Threshers

are Money-making Machines

because they are built of honest material by skilled, conscientious workmen, and have been proved to have the highest possible efficiency in doing the work for which they are intended.

Back of the SAWYER-MASSEY product is a strong organization, built up through seventy-seven years experience.

Sawyer-Massey 1913 Ontario Catalogue.

We have just issued the Ontario edition of our 1913 Catalogue, which illustrates and describes our full line of Portable and Traction Steam Engines, Gas Tractors, Threshers, Clover Hullers, Saw-Mills and road-making Machinery, with special reference to the requirements of Eastern Canada.

A copy of this book, handsomely illustrated in colors, will be sent on request.

Sawyer-Massey Company, Limited.

Builders of Steam and Gas Tractors, Threshers, Clover Hullers and Road-Making Machinery.

HAMILTON, ONT.

BRANCHES-WINNIPEG MAN. AND REGINA SASK.

This 1913 Sawyer-Massey catalogue was produced after the Massey family withdrew from the company in 1910. The advertisement offers gas tractors as well as steam tractors.

BULLETIN No. 323C

THE SAWYER-MASSEY

Canadian-Built

20-40 and 25-50

Kerosene-Burning Tractor

Sawyer-Massey produced their first gasoline tractor in 1910 as the Model 22-45. Several models were produced up to the mid-1920s. All but the smallest were designed for large area work as depicted here with a Sawyer-Massey tractor pulling a multi furrow gang plough.

Positive Kerosene Burning at All Loads

Has Already Made Good in the Field

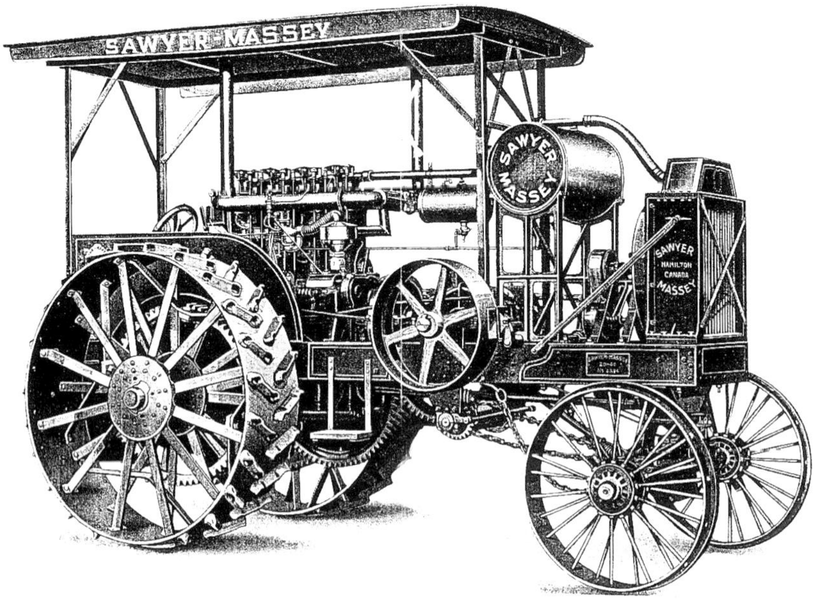

RIGHT SIDE OF SAWYER-MASSEY 20-40

Kerosene Burning Tractor Complete in Detail Ready for Operation

The 20-40 was typical of the Sawyer-Massey gasoline tractor style. It was built from 1918 to the mid-1920s.

Opening up Canada at the turn of the century required great lengths of roads and tracks to be laid. Sawyer and Massey produced a range of machinery for the task. 1900.

RIGHT SIDE VIEW SAWYER-MASSEY ROLLER

These Rollers are winners and wherever placed by SAWYER-MASSEY COMPANY, have won highest praise from the purchasers. The addresses of such, we will gladly furnish for confirmation of facts.

Their winning quality is the great compression effectiveness and the rapidity and consequent economy with which permanent and durable rolling is accomplished.

This is due to scientific design and correct distribution of weight, together with the great convenience of the working parts, which give the engineer complete control, at all times, of his Roller.

Included in these points of excellence is the high quality of material entered into the construction ; the careful, thorough and workmanlike manner in which every part is assembled and put together, ensuring easy and noiseless working of said parts and long life for which the Roller is celebrated.

To the thinking purchaser, it always appeals as the better investment, to pay the price for a good Roller, even though it be a few dollars more, when its durability can be proven double the life of the average, and where very little, if any, expense is incurred in its up-keep.

YOURS FOR VALUE,

SAWYER-MASSEY CO. LIMITED,

Good Roads Machine Department,

HAMILTON, CANADA.

References from the County road departments gave valuable promotional strength to Sawyer-Massey products. 1911

This 1895 catalogue front cover placed more emphasis on M-H seeding equipment.

When M-H purchased the Bain Wagon Co. of Woodstock, Ontario, in 1895 it gave them access to a remarkable range of wagons and sleighs for agricultural, forestry and general use. The Bain name was retained on the products, as shown on the rear and front cover of this full range brochure from the 1890s.

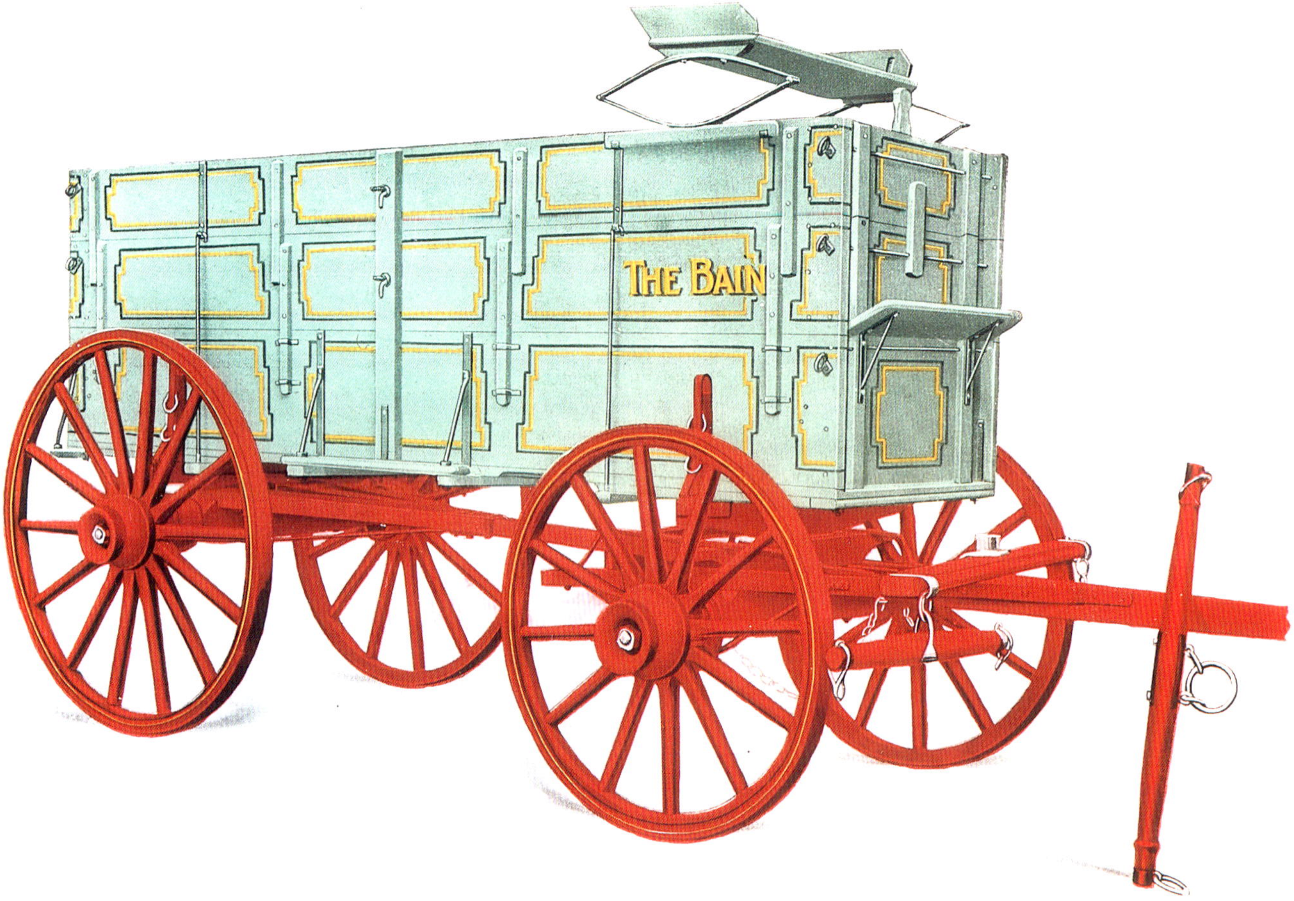

The Bain Triple Box Farm Wagon

Just one of the many specialist wagons from the Bain range.

1897 marked the 50th anniversary of Daniel Massey starting the company which later became Massey-Harris. A plentiful, global harvest is indicated to be the results of the company's efforts.

TO OUR EMPLOYEES
AT TORONTO
AND BRANTFORD:

Toronto, December 30th, 1897.

The Board of Directors of the Company being desirous of expressing their good wishes to the staff of Workmen by presenting each Employee with a New Year's Turkey, have found difficulty in securing such an immense number of Gobblers as are necessary at one time. They therefore leave each housewife to select her own bird—Turkey, Goose, Duck, or whatever suits her taste—with the brand new Dollar Bill enclosed. This bill is a new issue just recently brought out by the Dominion Government, and is graced with the Portraits of His Excellency the Rt. Hon. Earl of Aberdeen, Governor General of Canada, and Her Excellency the Countess of Aberdeen, which seems especially appropriate and interesting in view of their Excellencies' recent visit to our Toronto Works.

We sincerely wish you a bright, prosperous and happy New Year,

MASSEY-HARRIS CO., LIMITED,
W. E. H. MASSEY, *President.*

THE TORONTO MOWER No. 2 IN THE FIELD.

That Hart Massey was inventive in the use of publicity is shown by the use of his portrait on 3-cent "Massey Postage" in 1887.

M-H's 50th anniversary year saw the company showing great concern about the main course for their employees' New Year's dinner. Production of the Toronto mowers continued, and M-H President Hart Massey was inventive in his use of 'Massey Postage' in advertising to farmers everywhere. 1897.

In 1897, M-H introduced ball and roller bearings to replace bushes in binders. This dark coloured advertisement appeared in a 1901 M-H catalogue.

The front and rear covers of an 1898 M-H catalogue depict farmers as the civilizing influence on the world – aided by M-H exports of machinery creating an abundance of food.

The global domination of Canadian M-H is expressed by this 1900 catalogue.

This 1900 advertisement shows M-H highlighting the awards of excellence received for their machinery.

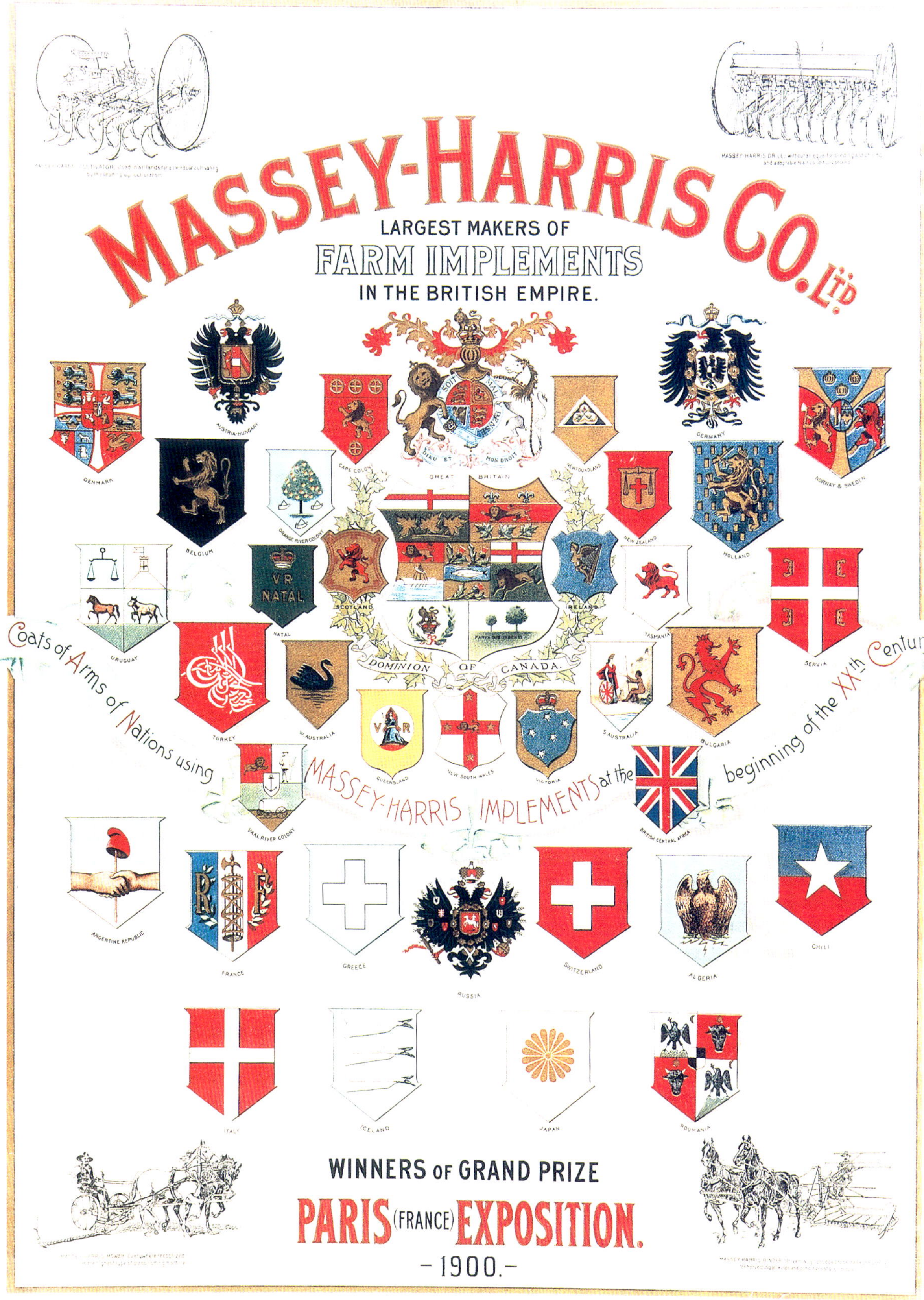

M-H was not slow to advertise its global trading success and the award winning merits of its products. 1901

A splendid emphasis is placed on M-H export strength, quality of product and the company's premier manufacturing position in the British Empire. 1901.

This Christmas 1903 M-H advertisement in the *Globe* magazine depicts a salesman of the day carrying a portfolio showing the wide range of equipment which could be purchased from M-H.

This 1905 advertisement shows that M-H Canada could supply its machines worldwide and year round from its extensive modern Toronto factory which was linked into Canada's rail network.

In 1907, the 60th anniversary year M-H catalogue cover assumed almost a modern environmental message with wildlife and wheat seen as mutual products of the countryside. The company's product portfolio in this 64-page catalogue included grain binders, corn harvesters, mowers, rakes, side delivery rakes, tedders, hay loaders, cultivators, seeders, hoe drills, disc drills, shoe drills, disc harrows, drag harrows, feed cutters, ensilage cutters, pulpers, manure spreaders, ploughs, wagons, sleighs and cream separators. 1907.

In the period c.1908–1910 M-H was claiming absolute supremacy for its equipment by using the term 'Monarchs'. An impression is also given of the way in which wide-scale adoption of their equipment enabled the cultivation and harvest of vast areas of wheat. The change in Prairie landscape is self-evident!

In 1910 M-H purchased the Johnston Harvester Co. of New York. This was M-H's first purchase of a non-Canadian company and it made M-H one of the largest implement manufacturers in the world. Johnston had great experience in reaper development and international marketing. This 1892/3 Johnston advertisement marks the 500th anniversary of Columbus in the 'New World'.

The luxurious artwork styles developed by M-H in Canada were to be used worldwide. This is the cover of a 1911 Hungarian M-H catalogue.

MASSEY-HARRIS
MASSEY-HARRIS
BUDAPEST,
V., Alkotmány-utcza 23.

For this 1913 catalogue, M-H and nature are depicted as natural partners.

A widening range of products is noted by this 1913 catalogue's rear cover, with emphasis being placed on the trademark guaranteeing quality.

M-H's first 'own' tractor came in 1914 when they formed a link with the Big Bull Co. to market their Big Bull tractors. Shown here is the centrepiece of a fold-out brochure of the tractor.
The subsoiling was achieved from the penetrating action of lugs on the wheels, but this is not shown here.

Massey-Harris Gasoline Engine—2 to 15 H.P.
Stationary, Portable or Semi-Portable

In 1911 M-H linked with the Seager Engine Works in Michigan to gain access to Olds stationary engines. This agreement lapsed and in 1913 M-H purchased the Deyo-Macey Co. who made engines in New York, and then in 1916 transferred production to their Toronto factory. A 1917 engine is shown here.

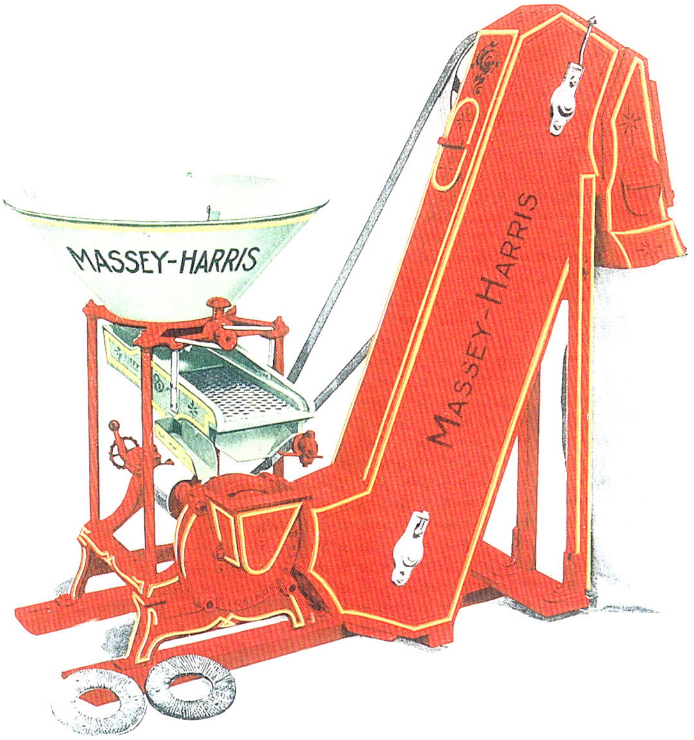

Massey-Harris Grinder—Three Styles—Seven Sizes

An extensive range of barn machinery was offered by M-H. c. 1917.

Massey-Harris No. 2 Tilting Frame Wood Saw

Great areas of forest were cleared to make new farming areas in parts of Canada, and of course Canada has always had an indigenous forestry industry. M-H were firmly into the timber machinery market. c. 1917.

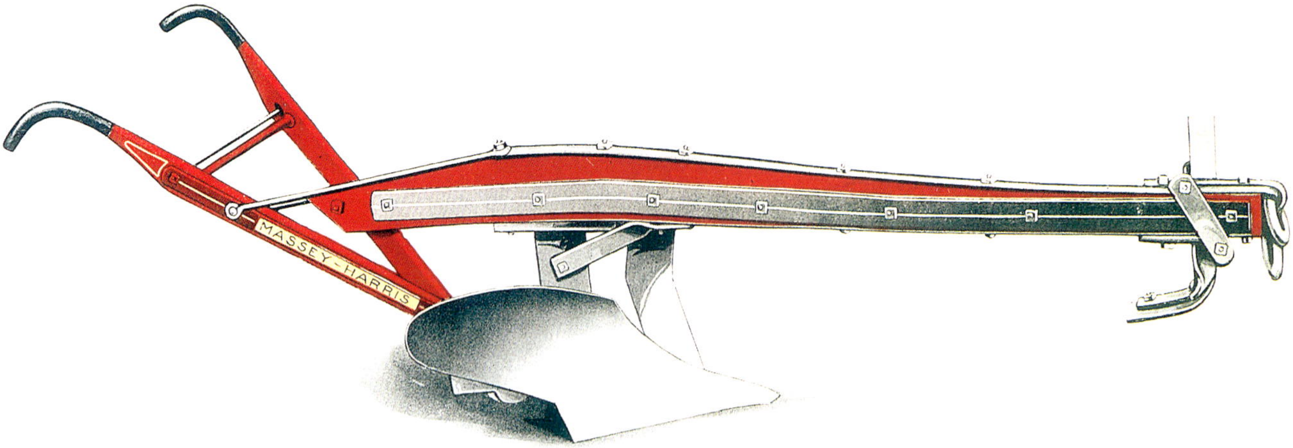

Massey-Harris No. 3 Railroad Grading Plow

M-H supplied ploughs to help with the railroad building effort in Canada. This advertisement can be dated to about 1917.

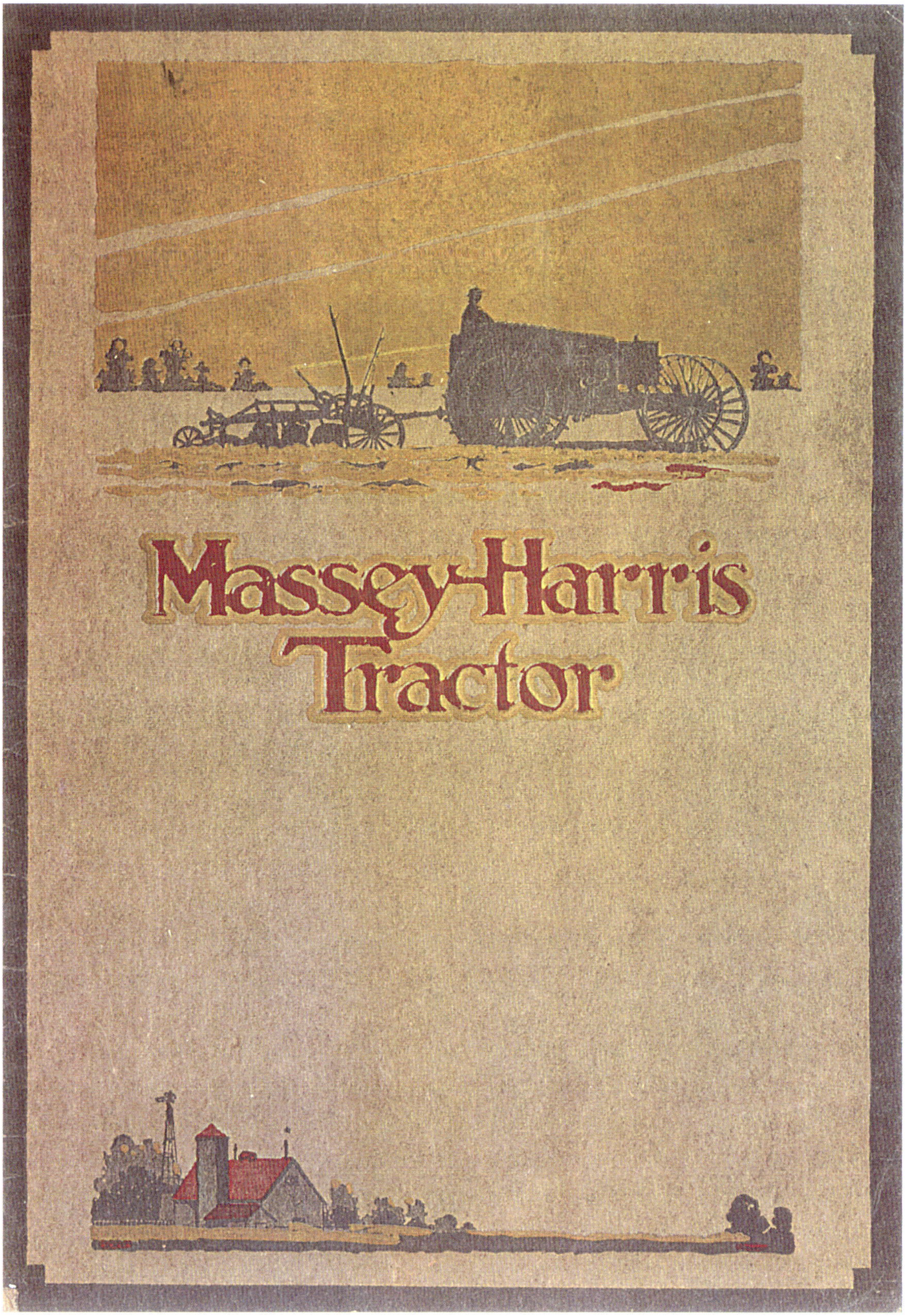

After dropping the Bull tractor, M-H went to Parretts for their tractor design, establishing a plant in Ontario in 1917 for their production. Three models were successively produced as the M-H No 1, No 2 and No 3 tractors. Production ceased in 1923. c. 1918.

An Aerial View of the Sunshine Harvester Works.
(Inset:—The Farm Smithy in which Hugh Victor McKay invented and built the first Sunshine Stripper Harvester.)

The amalgamation of H. V. McKay of Australia and M-H in 1930 marked a significant expansion of M-H's overseas activities. This 1931 photograph of McKay's Sunshine Works shows the size of the business.

Even before M-H amalgamated with H. V. McKay in 1930, M-H had been involved with McKay in combine harvester development. McKay were leaders in the field. 1901.

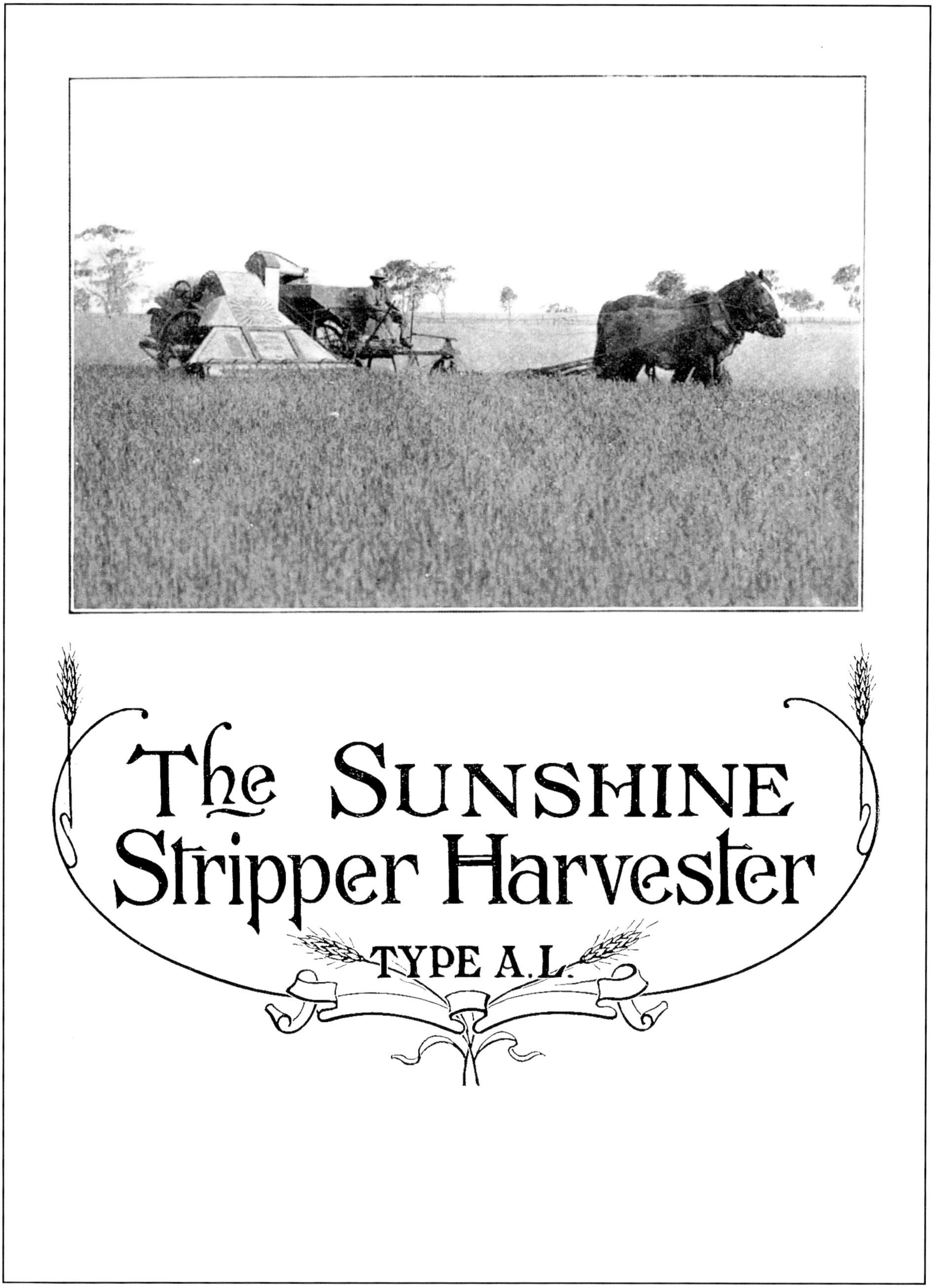

This H. V. McKay Sunshine Stripper Harvester with a 10ft cut pre-dates the amalgamation with M-H.

This M-H Reaper Thresher is thought to be from the 1900–1910 period and may be an early M-H machine developed in collaboration with H. V. McKay.

The Sunshine Chaffcutters.

FAVOURITES WITH THE FARMERS FOR TWENTY YEARS.

Unexcelled as a Chaff-cutting Machine for Farmers' use. Safe to Work, of Large Capacity, and built to last.

NO FARM SHOULD LACK A CHAFFCUTTER—A GOOD "SUNSHINE."

No. 2 X PORTABLE, WITH SHORT ELEVATOR AND BAGGING SPOUTS
(Cover removed, showing knives and gearing.)

H. V. McKAY PTY. LTD.,

SUNSHINE HARVESTER WORKS, SUNSHINE.

BRANCHES:

27-39 GEORGE STREET WEST, SYDNEY.
Telegrams, " Sunshine, Sydney."

CR. MURRAY & KING STREETS, PERTH.
" Telegrams, " Sunshine, Perth."

664 BOURKE-STREET WEST, MELBOURNE.
Telegrams, " Sunshine, Melbourne."

102-105 NORTH TERRACE, ADELAIDE.
Telegrams, " Sunshine, Adelaide."

DISTRIBUTORS IN QUEENSLAND:

H. V. MCKAY (QUEENSLAND) LTD., STANLEY STREET, South Brisbane. Telegrams, " Sunshine, Brisbane."

A separate Illustrated Catalogue of the Sunshine Engines will be posted to any address on request.

Sunshine Harvester Press.

Like M-H, H. V. McKay offered a wide range of machines including barn machinery. The bottom line shows that, like Massey, they also had their own publishing facilities.

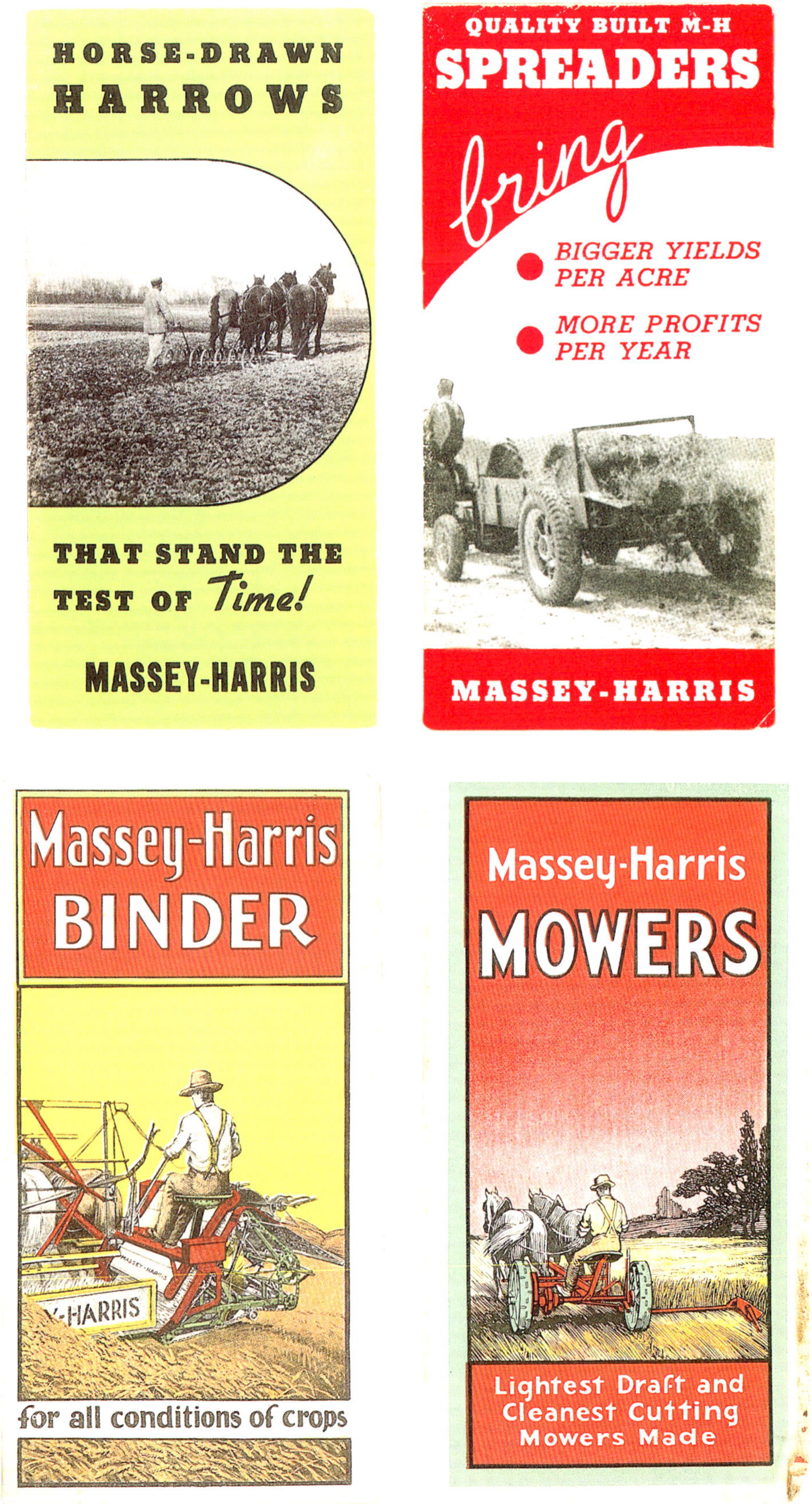

Over a long period of time M-H produced numerous small (3 x 6in.) fold out brochures to advertise their implements. They were displayed in racks in dealers' showrooms.

THE
New
MASSEY
HARRIS
MODEL
25
TRACTOR

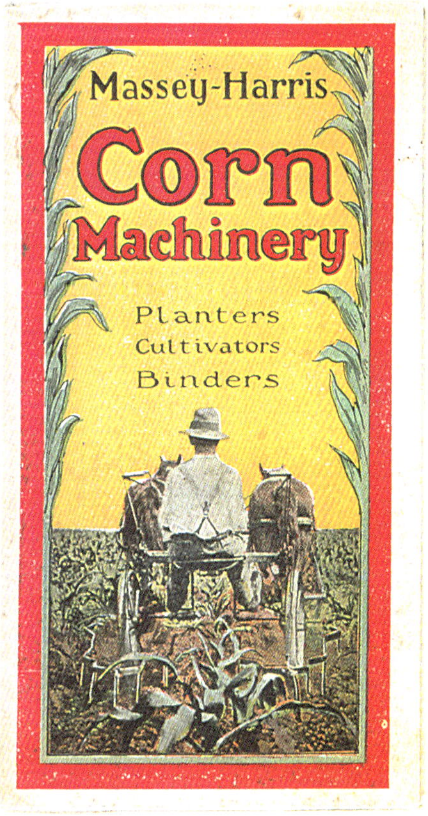
Massey-Harris
Corn
Machinery
Planters
Cultivators
Binders

MASSEY-HARRIS
DISC & DRAG
HARROWS
Make Good
Seed
Beds

MASSEY-HARRIS
GASOLINE OR KEROSENE
ENGINES
POWER IN A HANDY FORM

. . . when the
MASSEY-HARRIS
ROW CROP TRACTOR
COMES TO YOUR FARM

MASSEY-HARRIS
MANURE
SPREADERS

MASSEY-HARRIS
DRILLS
FOR UNIFORM
POSITIVE
SOWING
GRAIN DRILLS
FERTILIZER DRILLS
CORN PLANTERS

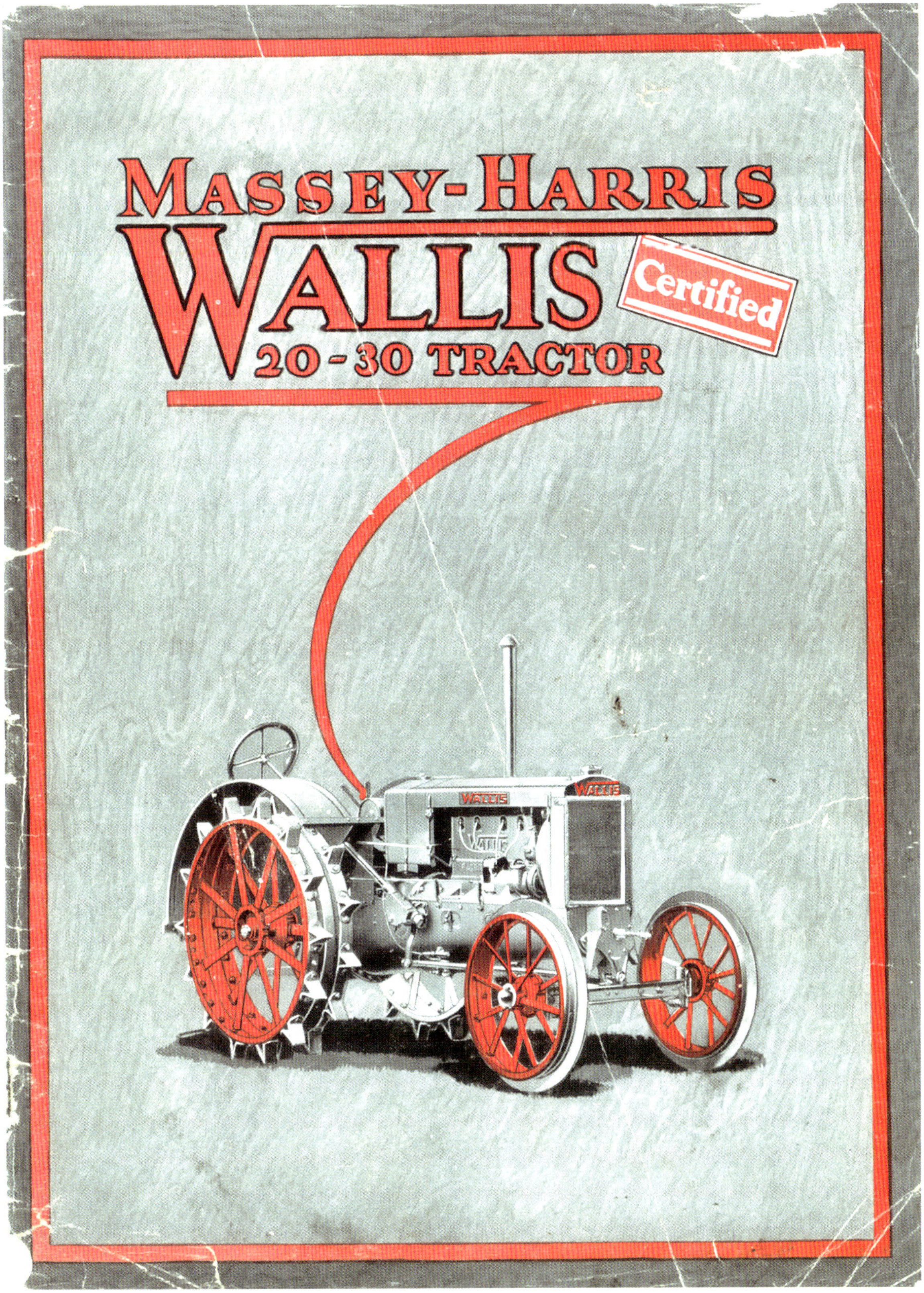

In 1928 M-H made its first truly successful move into the tractor market, a case of third time lucky, when it acquired the J. I. Case Plow Co., which made Wallis tractors. Seen here is the principal tractor that came with the deal – the Wallis 20-30 – soon to be redesignated the M-H 20-30.

1930 saw M-H release its first own-design tractor – the M-H General Purpose 4WD. It was ahead of its time and reportedly had mechanical problems. The 1930 model had an OHV Hercules engine whilst a later 1936 model had an M-H engine.

Sundex Farm Engines.

Sizes 2 H.P. and 4 H.P.

Illustrating the Sundex 2 H.P.

Horizontal. Open Crankcase. Hopper Cooled. 4-Cycle. Volume Governed. Ignition by High Tension Magneto. Renewable Sleeve Cylinder.

Thousands of these finely finished Engines are in use, year in and year out, driving Chaff-cutters, Saws, Pumping Plants, Milking Machines, Separators, &c., justifying the confidence of the owners by their Efficiency, Economy of Petrol Consumption, and Lasting Reliability.

THE SUNDEX IS DESIGNED ON SIMPLE LINES, SO THAT A NOVICE QUICKLY BECOMES CONVERSANT WITH ITS ACTION AND THE LITTLE ATTENTION THAT IS REQUIRED TO ENSURE SWEET RUNNING.

The Crankcase is of heavy design, forming a solid foundation for the Engine. The crankshaft bearings are of large dimensions and are accessible. Stout white metal bearings are provided for these journals, which can be replaced, when worn, at a minimum of work and cost. They are each lubricated by pressure grease cups of durable character. The crankshaft is a one-piece drop forging of steel of vast strength at the throw and at the bearings. The connecting rod is also of steel, carefully and accurately machined.

The Cylinder is an independent sleeve, driven into position in the main casting, and it may also be replaced by a new one after years of wear. The cylinder walls are ground to a fine degree of accuracy and to a glassy smoothness by special machinery.

H. V. McKAY MASSEY HARRIS PTY. LTD.
SUNSHINE HARVESTER WORKS, SUNSHINE.

H. V. McKay of Australia also marketed stationary engines. 1931.

Ploughs, ploughs and more ploughs for tractors and horses is the M-H message here – spread over the rear and front covers of a brochure thought to have been produced in about 1931.

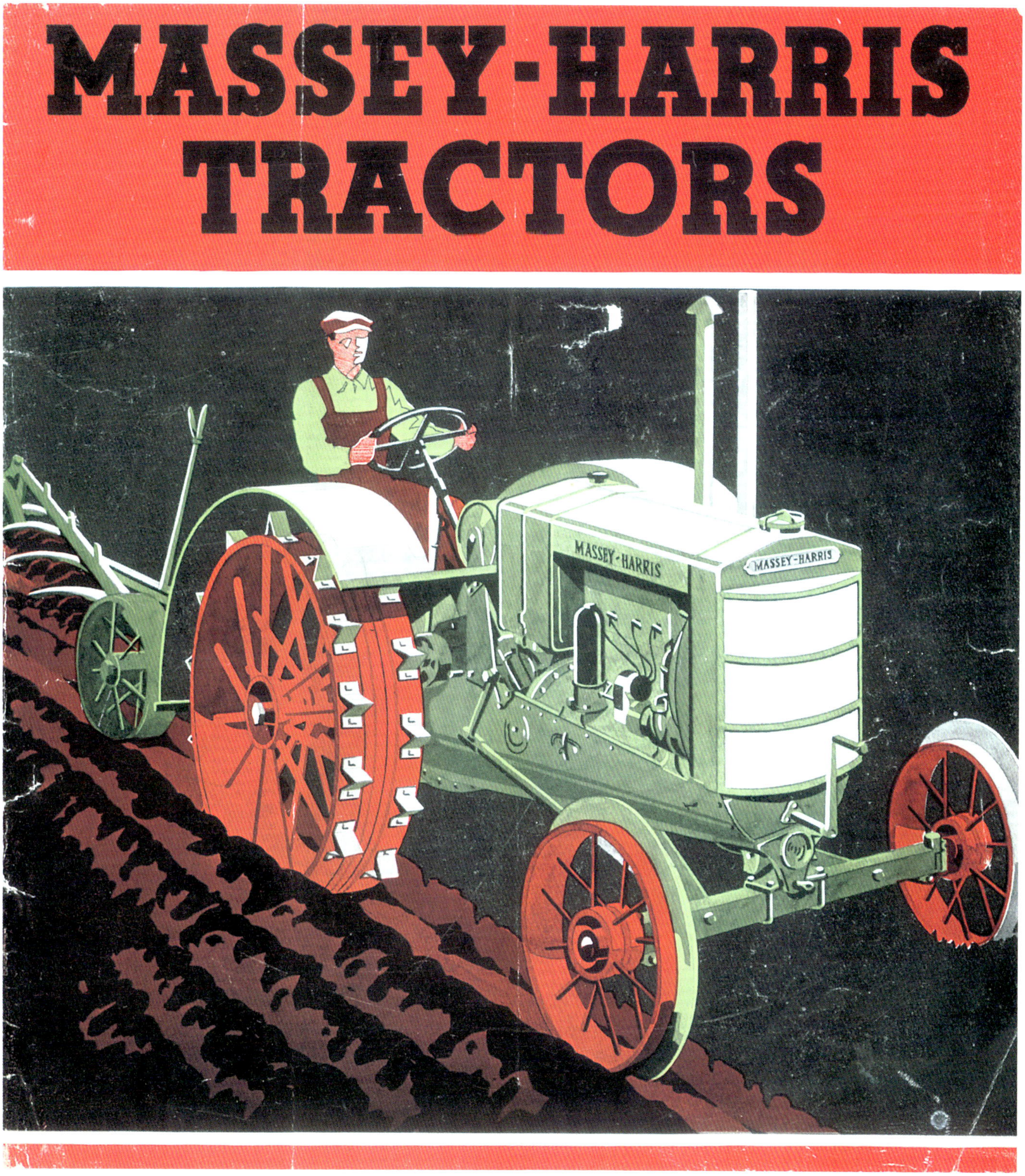

This 1930s UK M-H Tractors catalogue features a green M-H 25-40 tractor on the front cover. The 25-40 was introduced in 1931. Other tractors featured were the 12-20, 20-30 and 4WD General Purpose.

The *Farmers Handy Catalogue* was a long running series of Canadian M-H catalogues. They are extremely comprehensive and well worth acquisition by any M-H enthusiast. This 1937 90th anniversary edition shows 90 years of progress as essentially being from sickle to combine, while an idyllic rural setting is maintained. As far as can be established, this is the first *Farmers Handy Catalogue*.

Cushman was a later type of stationary engine to be marketed by M-H. This advertisement comes from 1936.

M-H combine development proceeded rapidly in the 1920s and 1930s with a series of pull type models. This No 14 was first produced in 1936 as a 12ft or 16ft cut machine. Two years later M-H produced their first self-propelled machine, the M-H No 20. 1937.

As this 1937 advertisement shows, George White threshing machines were badged Massey-Harris.

A 1939 UK M-H seed drill catalogue with prices ranging from £36.5.0 to £110.0.0.

In 1933 Blackstone and Co. agreed to manufacture exclusively for M-H in the UK. 1938.

This small 4.75 x 6in. 1938 UK Massey-Harris Blackstone catalogue is superbly illustrated with almost fluorescent artwork for the inside pictures.

Elevators were a common sight on British farms but apparently not so in North America. The advertisement comes from 1938.

The Blackstone lines of equipment were featured exclusively in these small (3.5 x 5.5in.) brochures.

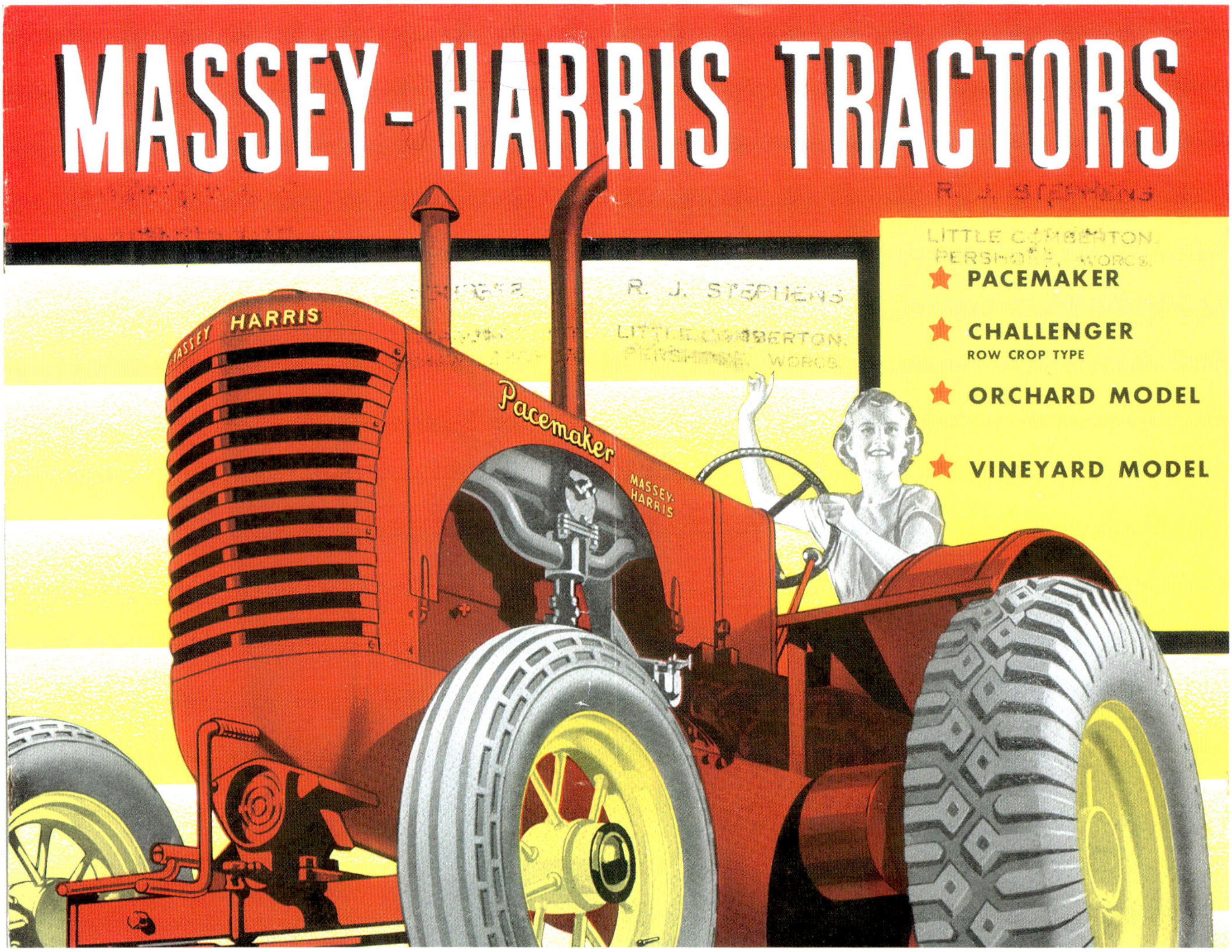

This UK brochure boldly advertises one of the last of the Wallis style tractors – the Red Pacemaker. Very few of the particular gasoline model shown here were sold in the UK due to the price of gasoline. This 1938/9 picture may well be one of the first in which M-H showed women at work in their advertising.

Romantic British countryside is shown here in order to enhance the work and mechanical beauty of a horse-drawn binder. It is overprinted with a South Wales M-H agent's name and address.

North American-made disc harrows of this type were a strong selling line in the UK in the 1940s.

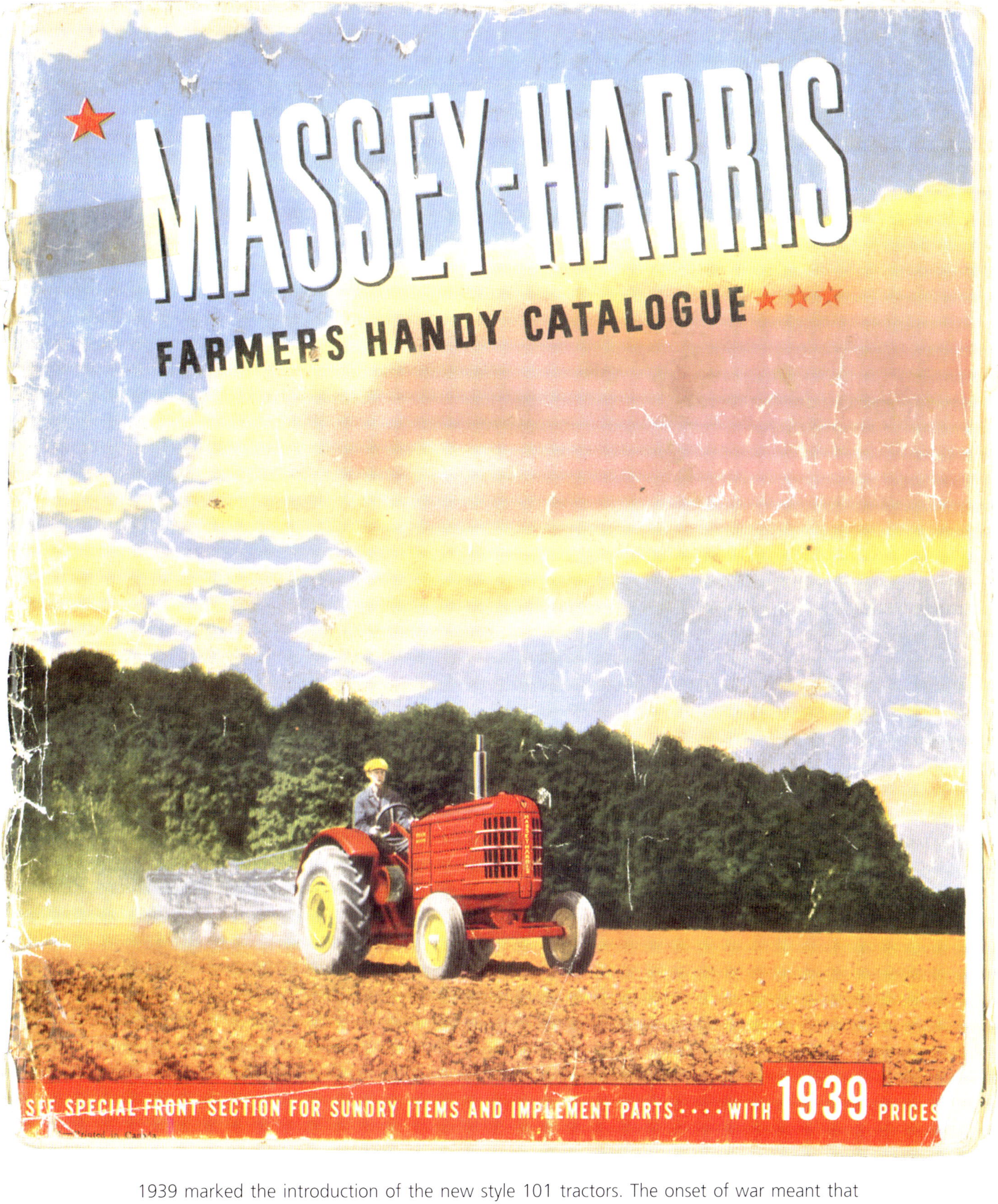

1939 marked the introduction of the new style 101 tractors. The onset of war meant that tractors had much hess steel. In this particular Canadian catalogue it is advertised alongside remaining stocks of red Pacemaker and 25 tractors.

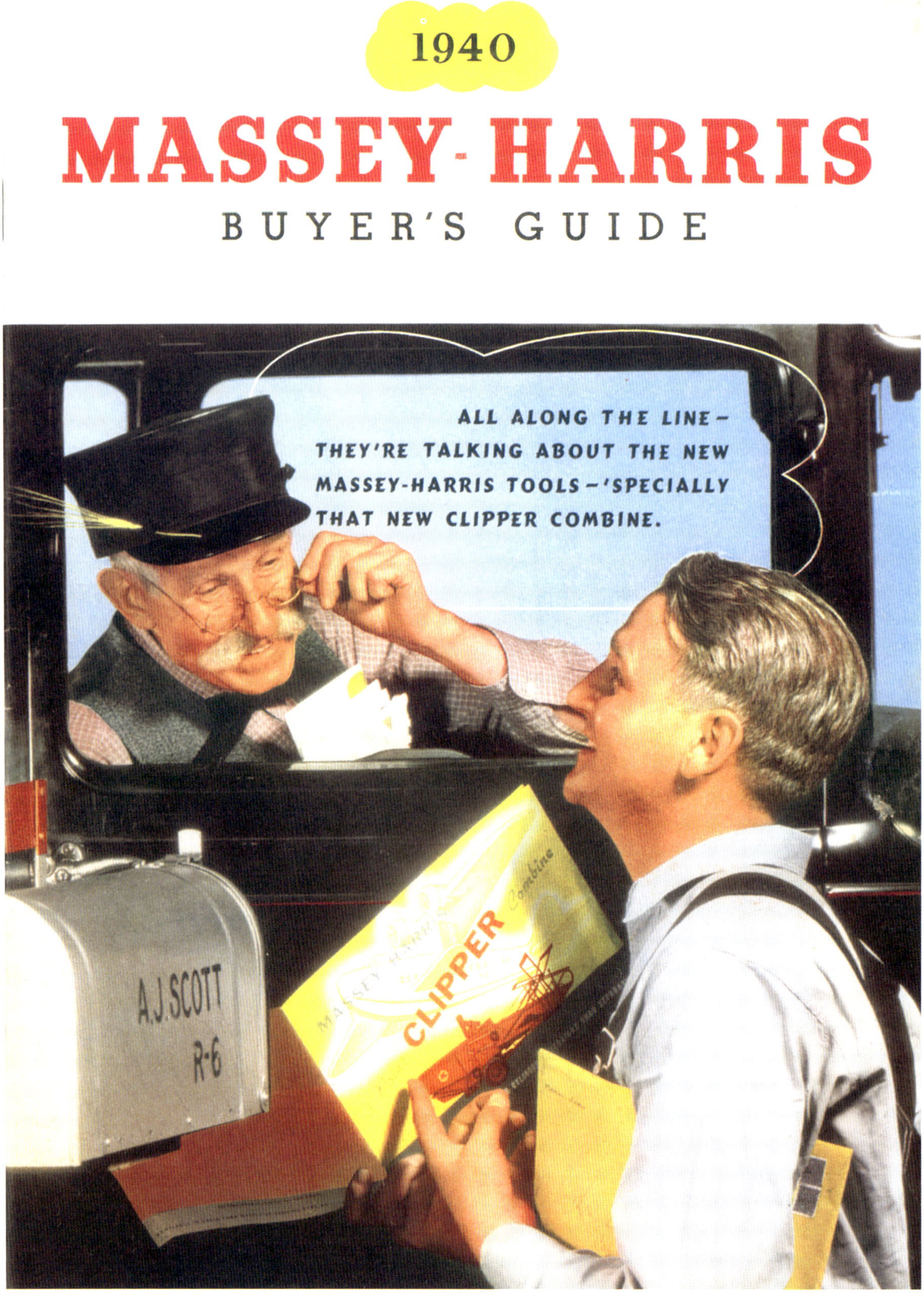

In the USA, the *Massey-Harris Buyer's Guide* was equivalent to the Canadian *Farmers Handy Catalogue*. However, being an 'overseas' market, the USA Buyers Guides contained far fewer items. Here we see the importance of postal advertising to the company.

The 101 Junior and its successor, the 102 Junior, were one of M-Hís best selling wartime tractor models. The driver's helmet may indicate part of M-Hís contribution to the war effort.

Massey Harris sold the General tractor, made by Cleveland, from about 1942 to 1946. Some reports suggest it was marketed earlier. It was possibly used to test the small tractor market before the M-H Pony tractor was produced. The General had a Hercules engine and the tractor had a 10–17hp rating.

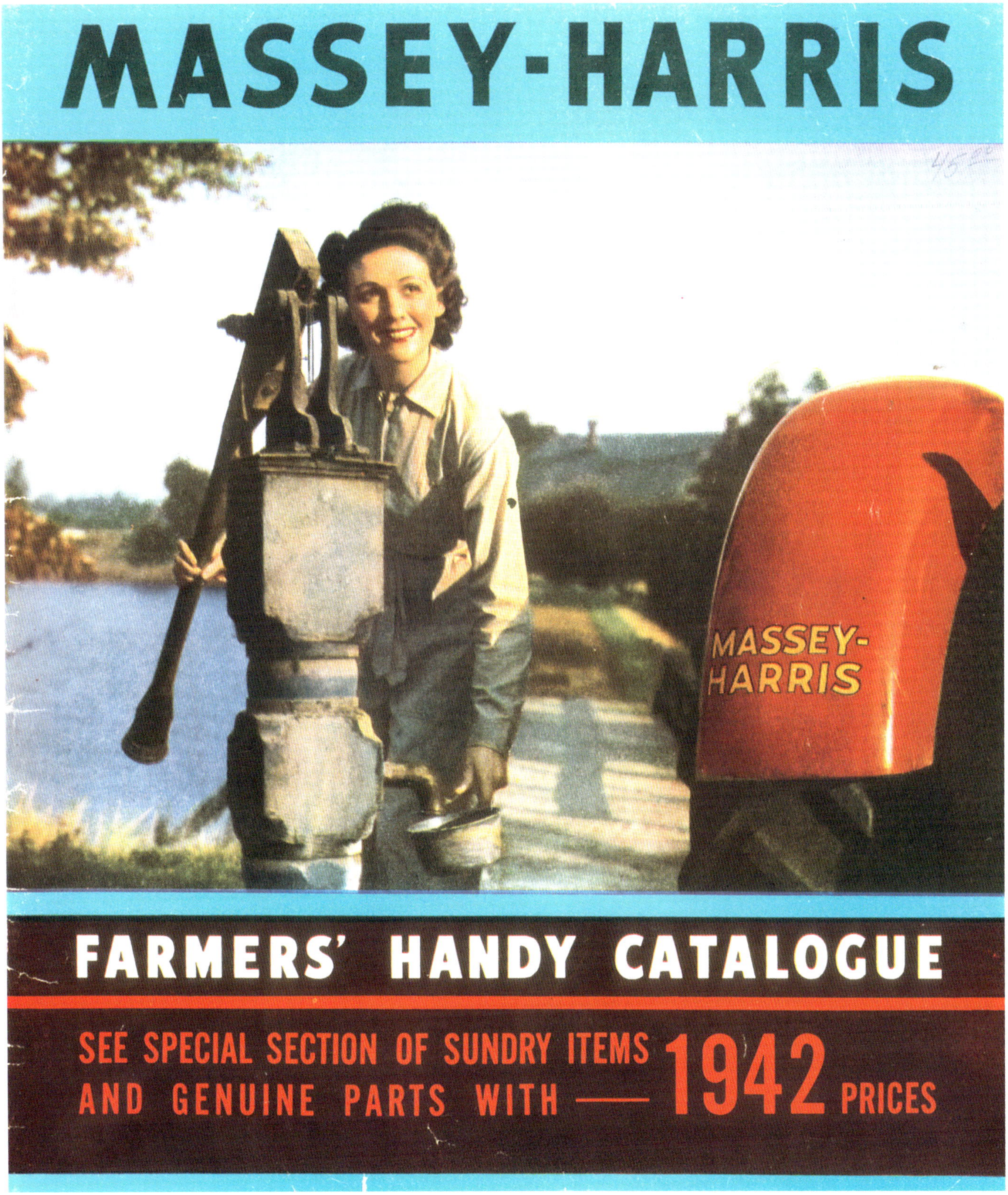

The middle of World War II saw many women across the globe taking up farm work. 81, 82, 101 Junior, 102 Junior, 101 Senior, 102 Senior, 203, 202 and 201 tractors, as well as the small General tractor, were available in this 1942 catalogue.

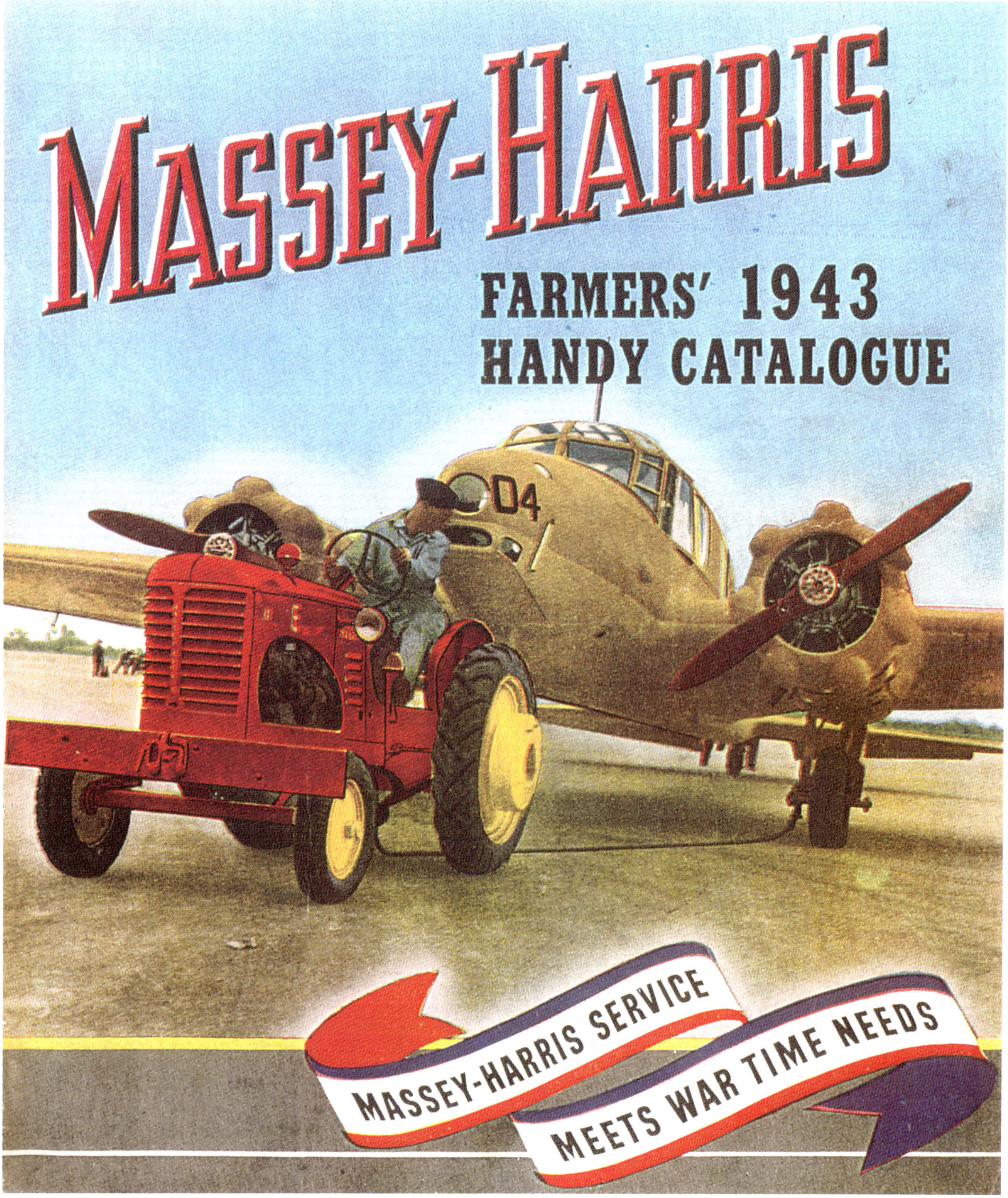

M-H never made aeroplanes but the M-H 81 '2 plow' tractor was widely used for airfield work by the Royal Canadian Air Force.

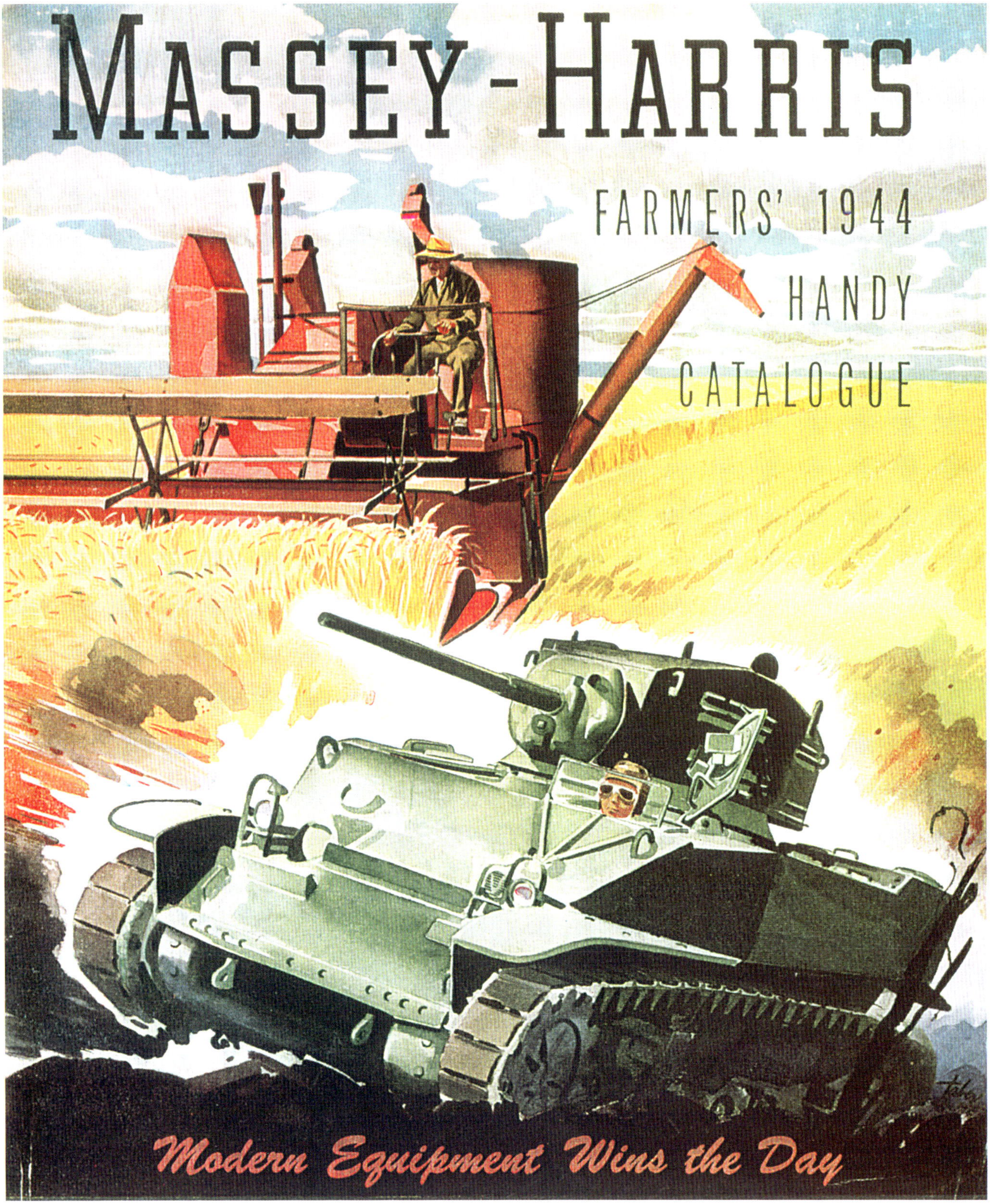

M-H turned factory facilities over to armaments manufacture during the war. Although M-H produced significant numbers of tanks at their Racine factory, they were not advertised for sale in this catalogue! But M-H does publicize its contribution to the war effort via both agricultural machinery and armaments production.

The M-H-inspired Harvest Brigade gave the company unique publicity for its self-propelled combine harvester, as well as to the very real and tangible goal which these machines achieved.

Just as M-H tanks were self-propelled, so were M-H combine harvesters. Both were pushing for victory in World War II.

M-H reached their 100th anniversary only six years before merging with Ferguson. The anniversary logo lays emphasis on M-H's long history of revolutionizing grain harvesting.

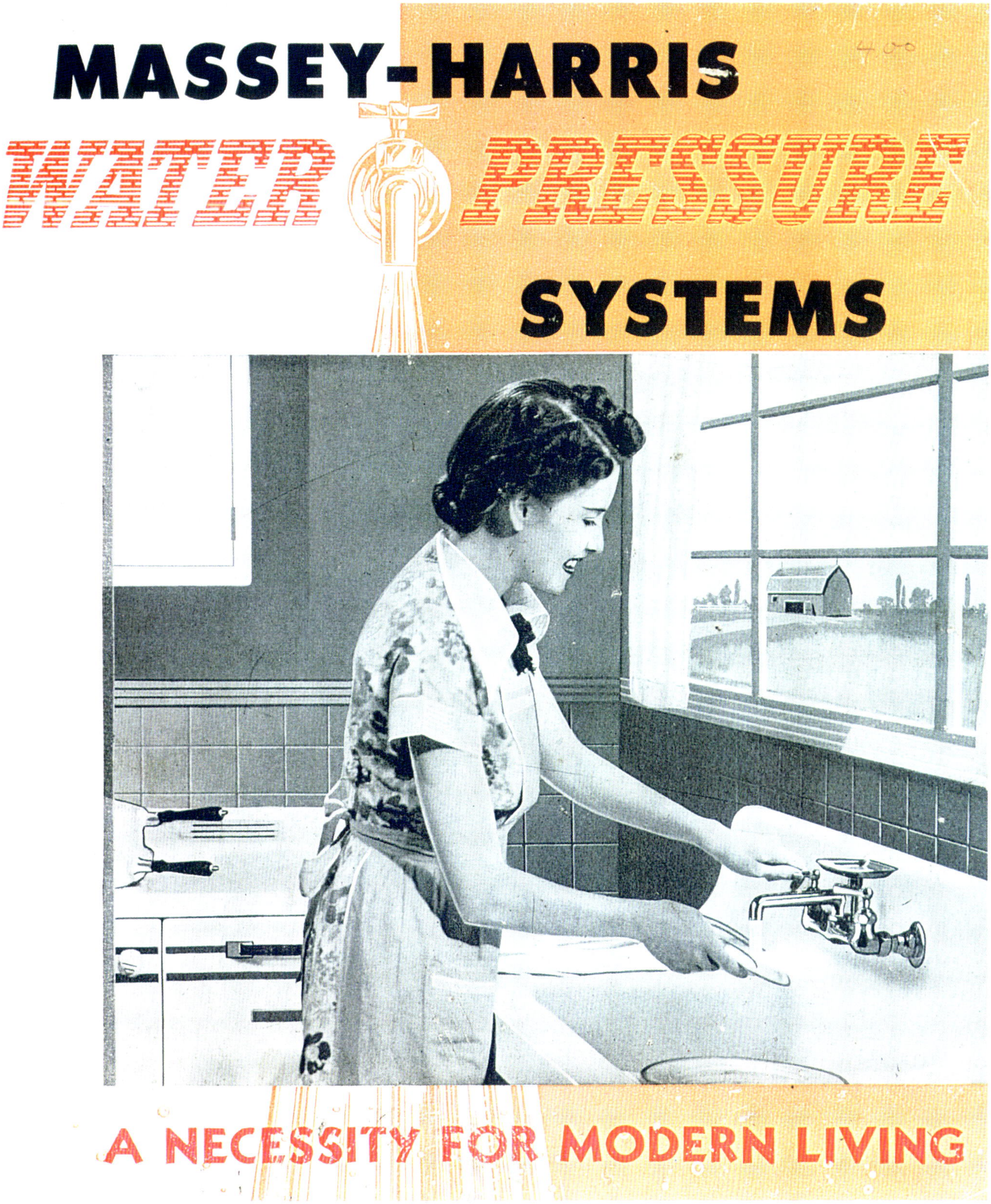

M-H responded to post-war demands for higher living standards in many ways. One of the most basic farm house requirements was a pressure water system. M-H pumps for this purpose were made by the Beatty Co. 1948.

A calm family farm scene for this 1948 catalogue which lists the new post-war range of tractors from the Pony up to the massive 55.

A 1948 air-cooled engine offered by M-H for driving their milking machines.

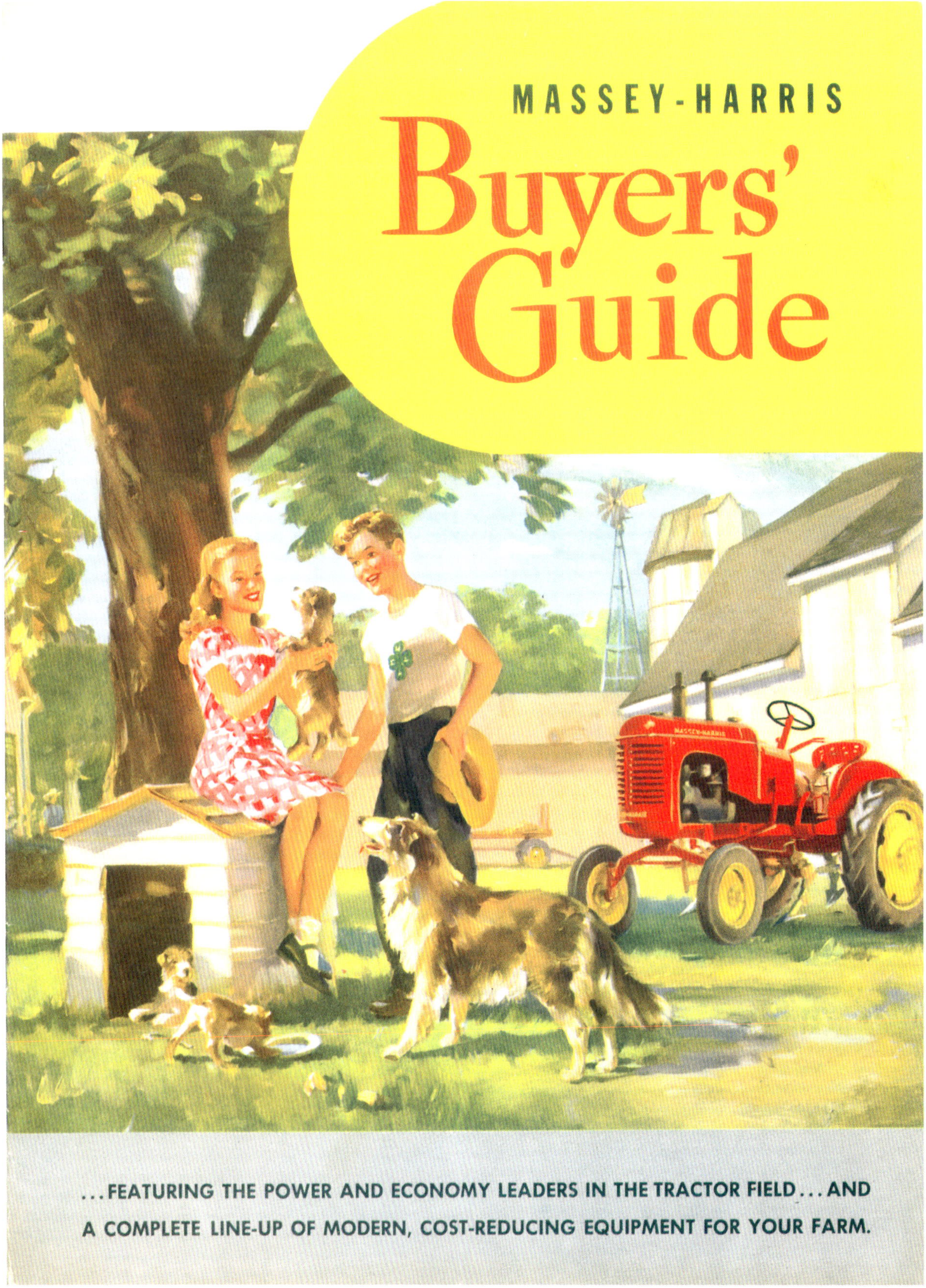

Pure enjoyment of life for children on an M-H equipped farm in 1949.

In Canada, M-H had traditionally marketed a wide range of products to its rural customers who were often in very isolated situations. By 1951 prosperity was returning after the ravages of war and M-H were offering washing machines. But horse-drawn ploughs and scufflers were still offered.

This 1951 USA M-H dealers' calendar, besides having seasonal pictures featuring some M-H equipment, also has useful tables and information on the rear.

The 1951 *Buyer's Guide* gives an impression of the dawning of the age of larger machinery dealers' showrooms. By 1951 the 26 and 27 self-propelled combine harvesters, and the self-propelled and trailed Clipper combines were taking over from older trailed combine models such as the Nos 15 and 17 which were still offered here. Combines were becoming more adaptable with the offering of tracks and rice tyres. A self-propelled corn picker and trailed forage clipper were also offered.

A family farm emphasis again for M-H in 1953, perhaps with a suggestion of M-H serving three generations.

Yes, it really is a *grey* Pony tractor! Most were red, but when M-H merged with Ferguson some were painted grey, both to appease Ferguson dealers and keep Pony sales moving in the face of the superior Ferguson tractors. 1954.

Four post-war M-H tractors bore 'horsy' names – the Pony, Pacer, Colt and Mustang. The Pacer and its implements are described in this 12-page brochure from 1956.

The M-H 50 was the first M-H badged tractor to have the Ferguson hydraulic system. It was of course a Ferguson 40 in M-H livery. This 3 x 5in. 30-page brochure depicted the tractor in comic strip form.

Dust bowl conditions in the Prairies triggered the development of tillage machinery which would conserve soil from both wind and rain erosion. This resulted in a shift from plough-based cultivation to reduced and minimal cultivation. One M-H response was the Trash King cultivator. The advertisement comes from 1953.

MASSEY-HARRIS
MULTI-PURPOSE VISE

The Handiest Tool
in the
Farm Workshop

M-H in Canada sold a wide range of tools and workshop supplies. One classic item was the M-H vise – a prized collector's item today.

M-H Buyer's Guides continued into the Massey Ferguson period. This one is possibly from 1957 and lists the 333, 444 and 555 tractors along with the Ferguson design M-H 50. The 60, 82 and 90 combines, including Hillside models, were available in this catalogue.

Silage making was well established in North America well before it was common practice in Europe. The M-H forage clipper was a widely used machine; it was pto or mounted engine driven, and with adjustable chop length. c. 1951.

The UK produced M-H 726 combine was the UK's first self-propelled combine to achieve a mass market. (The 7 denotes UK manufacture of M-H machines.)

The M-H 780 self-propelled combine was a stable mate to the 726. This 1952 painting is by the celebrated transport artist Terence Cuneo. Cuneo painted in the order of 20 such works on an M-H commission.

The M-H Electromatic Hammer Mill appears to have been a UK M-H product. The cleanliness of the worker suggests that milling was a dust-free activity! 1956.

After World War II, M-H's association with Blackstone came to an end and a new relationship was forged with James Dickie and Co. of Ayrshire, Scotland, for supply of UK-built hay making and other equipment. This relationship lasted into the early Massey-Harris-Ferguson era. This Dickie rake had a concrete balance weight at the rear. 1954.

MASSEY-HARRIS

NEW PATENTED COMBINED SWATH TURNER AND HAY COLLECTOR

Dickie

An Invaluable Tool for the Modern Hayfield

The new Massey-Harris Dickie patented combined Swath Turner and Hay Collector embodies exclusive features in design and performance, that place it in the front rank of its class.

Years of experience by specialists have at last produced a Swath Turner that will stand up to the stresses and strains of fast tractor work, whilst the wider sweep and higher speed of the turning heads ensure first class work at high tractor speeds. Leaves the swaths fluffed up loosely on the stubble, allowing sun and air to dry them out quickly and thoroughly.

Illustration shows the one-man tractor model with forward controls easily accessible from the tractor seat

The technical details outlined below will convince you that here is a Better Swath Turner—a machine that will be Indispensable to you at Haytime.

Read these special features :—

★ Trussed Fabricated Steel Frame Construction ★ Patented Automatic Shift for the Centre Screen, eliminating the Necessity for a Control Lever ★ Tines can be lifted In or Out of Work by "Finger-Tip" Control ★ Turning Heads Independently Adjustable for Swath Width and Rotation of Turning ★ Gear Shift Levers Quick and Easy to Operate ★ Turning Heads and Screens Quickly Contracted to Wheel Width for Transport ★ Works Equally Well on Ridge or Furrow ★ Safe on Hillsides ★ Tractor Model is One-man Outfit, Forward Controls Being Easily Accessible from Tractor Seat ★ Perfectly Balanced ★ Exceptionally Light Draught for Work with Horses Easy on Horse and Driver.

MASSEY-HARRIS LTD. Massey House, Brooklands Road, Sale, MANCHESTER

LIST 9401/11/45/B49181

This early version of the M-H Dickie swath turner had a channel iron frame; later ones had a tubular frame. A one (central) wheel version was also made for use on tractors with three-point linkage.

A *NEW* 14ft. TRACTOR RAKE

THE MASSEY-HARRIS DICKIE No.716

SPECIFICATION

Transport Width - - -	8′ 4″
Working Width - - - -	14′ 2″
Raking Width - - - -	14′ 0″
Weight - - - - - -	11 cwt.
Number of Tines - - -	48

** Massey-Harris reserves the right to make changes to design and specification without obligation to subsequent purchasers.*

The M-H Dickie 716 folding tractor rake with pneumatic wheels was not widely sold and is a rarely seen implement today.

Unless you have visited the central North American plains, you cannot appreciate the vastness and loneliness of the land areas which are farmed there. M-H had no doubts that they were up to the task. 1952.

M-H recognized the unique superiority of the Ferguson Hydraulic system long before the merger of the two companies. M-H made various hydraulic lifts, often called Depth-O-Matic, which were inferior to the Ferguson and even sponsored the University of Toronto to come up with a competitive design to the Ferguson. However, the project was abandoned as a failure in 1953 – the year of the merger with Ferguson.

The No 1 Slicer Hay Baler was M-H's first pick up baler which, they claimed, minimized leaf loss. The eight-page brochure is coloured green throughout to convey the impression of hay harvested in prime condition – not bleached by excess sun or rain.

The M-H Raussendorf Straw Press was sold on the European market. In essence it was an interim technology prior to the wide-scale adoption of pick up balers. The device produced low density bales, and is shown here on an M-H 780 combine.

M-H made self-propelled, tractor-mounted and trailed corn pickers. The self-propelled and mounted machines were two row, and the self-propelled model had a variable speed transmission with 24 speeds.

MASSEY-HARRIS

No. 721A HAMMER MILL

For effective, economical milling of grain or roughage

- ★ **Strong Steel Body**
- ★ **24 Reversible Hammers**
- ★ **Large Feed Hopper— Adjustable for height**
- ★ **Low Speed— Large Capacity**
- ★ **Heavy-Duty Ball Bearings**
- ★ **Positive Worm Feed— No clogging**

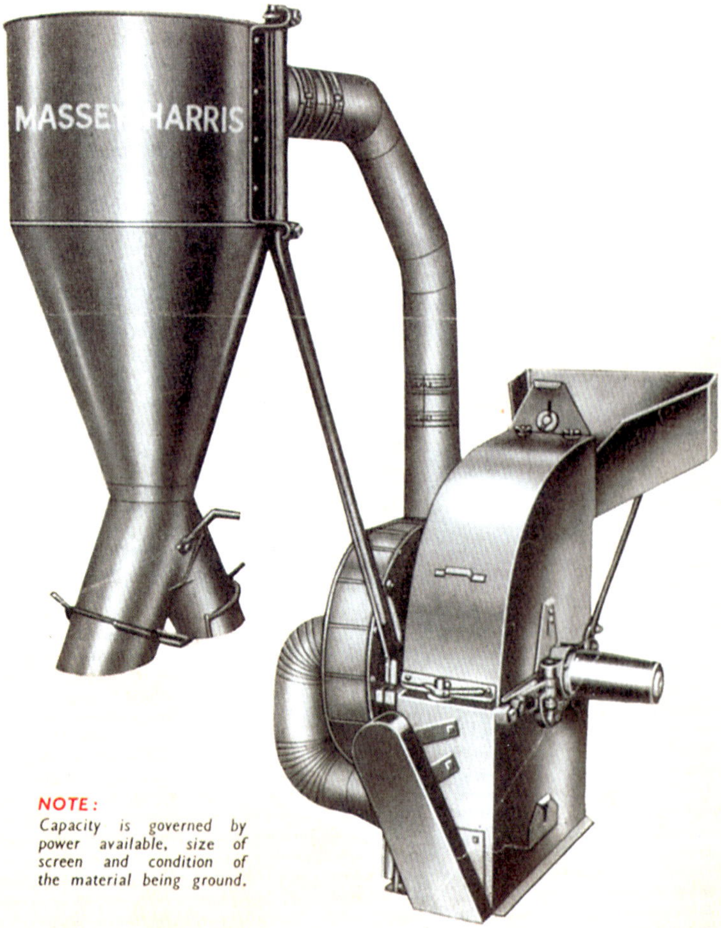

Approximate Capacity at 2,000 r.p.m.

Material	*Screen*	*Lbs. per hour*
Oats	1/4" 5/16" 3/8"	2,000 to 3,400 2,500 to 3,800 3,000 to 5,000
Barley and Wheat	1/8" 3/16" 1/4"	2,400 to 3,000 2,800 to 3,400 3,000 to 4,800
Sheaves	3/4" 1 1/2"	1,500 to 3,000 2,000 to 3,500

NOTE: Capacity is governed by power available, size of screen and condition of the material being ground.

MASSEY-HARRIS LIMITED *Pioneers of Farm Mechanization* **LONDON · MANCHESTER · KILMARNOCK**

The 721 hammer mill was a best-selling M-H line in the UK. It was assembled at the Manchester M-H factories.

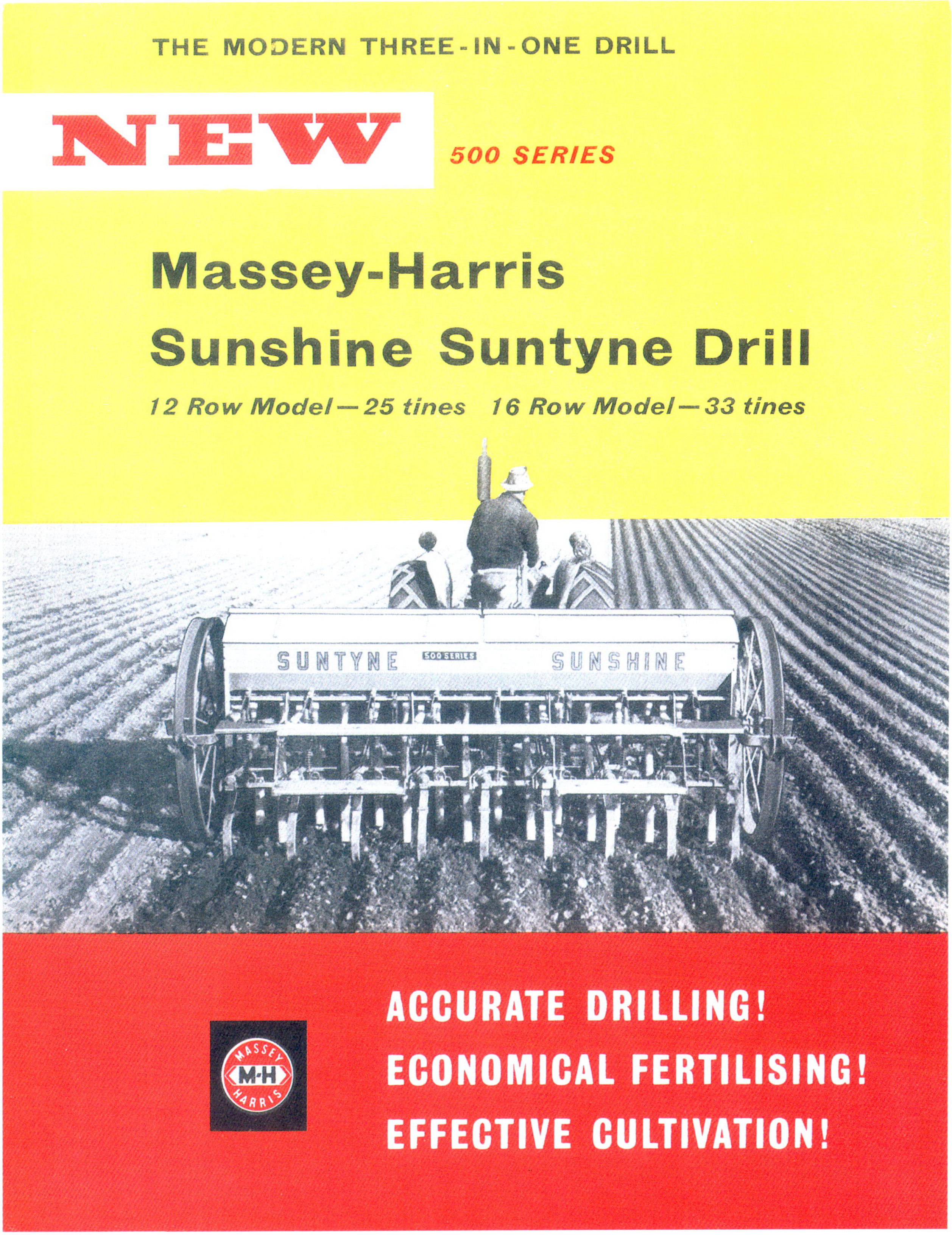

Australian M-H 'Sunshine' equipment was imported to the UK in significant quantities in World War II, and imports continued after the war. This model was still being sold in the M-H-F period. The advertisement comes from 1955.

This 1959 Australian brochure shows the M-H name still being used even though the company's name had changed to Massey Ferguson. The versatility of the Australian drills is evident from their variable design and adaptability to very varying soil situations.

Section IV

Ferguson 1917–1953

Harry Ferguson must be credited with being the agricultural engineer of the twentieth century. He revolutionized tractor design by recognizing that the traditional hitching of trailed cultivation implements to tractors by a single-point hitch was fundamentally flawed. He identified the fact that for mechanical efficiency, and improved operator safety, implements had to become an integral part of a tractor, and in operation transfer weight to the tractor rear wheels to aid traction.

Following a series of trial designs, the first implement which he marketed incorporating these principles – in 1917 – was a plough for an Eros agricultural conversion of a Ford model T car. Limited sales were to be achieved, but events were taken over by the development of the Fordson model F tractor. Ferguson recognized this as a practical lightweight tractor and adapted his plough design accordingly. Ferguson first had ploughs made for Fordson tractors in 1923 and 1924 by Roderic Lean in the USA, and then subsequently by Uber and George Sherman, also in the USA.

The plough linkage design progressively evolved from a two-point to a three-point linkage in the Fordson era. Subsequently the linkage achieved its sophisticated and hydraulically operated design with the production of Ferguson's prototype tractor in 1933 – the Ferguson Black tractor. This was commercialized in 1936 with the production of Ferguson Brown tractors during the years 1936–1939.

Ferguson then established a relationship with Henry Ford for the production of the Ford Ferguson tractor which was produced between 1939 and 1947. When this relationship failed, Ferguson set up his own manufacturing facilities, first in the UK at Coventry for the Ferguson TE series of tractors which ran from 1946 to 1956, and then in the USA at Detroit for the Ferguson TO series of tractors which ran from 1948.

The Ford Ferguson era saw a massive expansion in the range of implements offered for Ferguson tractors – only a very limited range had been offered with the Ferguson Brown. This was the world's first range of three-point linkage mounted, automatic hydraulic depth controlled cultivation implements, complemented by other types of mounted implements.

In 1953 an ageing and tiring Harry Ferguson merged his companies with Massey-Harris to form Massey-Harris-Ferguson. Visually the 306,221 Ford Fergusons, 215,000 American Fergusons, the 513,651 British Fergusons and the 74,655 grey Ferguson 35 tractors were very similar. Only the 1,250 Ferguson Browns produced at the start of the Ferguson tractor era, and the 9,097 Ferguson 40 tractors produced at the end were substantially very different to the 1,109,527 others. What makes the Ferguson era so interesting in advertising terms is perhaps not the tractors, but rather the great range of Ferguson implements designed for the Ferguson tractors. Efficient operation of the tractor was largely dependent on the use of these specific implements because its light weight required weight transfer from implement to tractor.

Early Ferguson plough development

This 1926 advertisement shows a Fordson tractor with a Ferguson Sherman plough. It is implicit that the plough sold the tractor – such was the unique and effective nature of the Ferguson mounted plough concept. At this stage of the evolution of the Ferguson system, the plough was the only implement offered. It was to be another ten years before an extended line of Ferguson style implements were offered with the Ferguson Brown tractor.

Advertising the many benefits of the Ferguson plough.

Total convenience with a Ferguson plough!

Ferguson Brown

This Ferguson Brown advertisement is probably not as honest as it might be. Many other makes of tractor were replacing horses well before the arrival of the Ferguson Brown tractor.

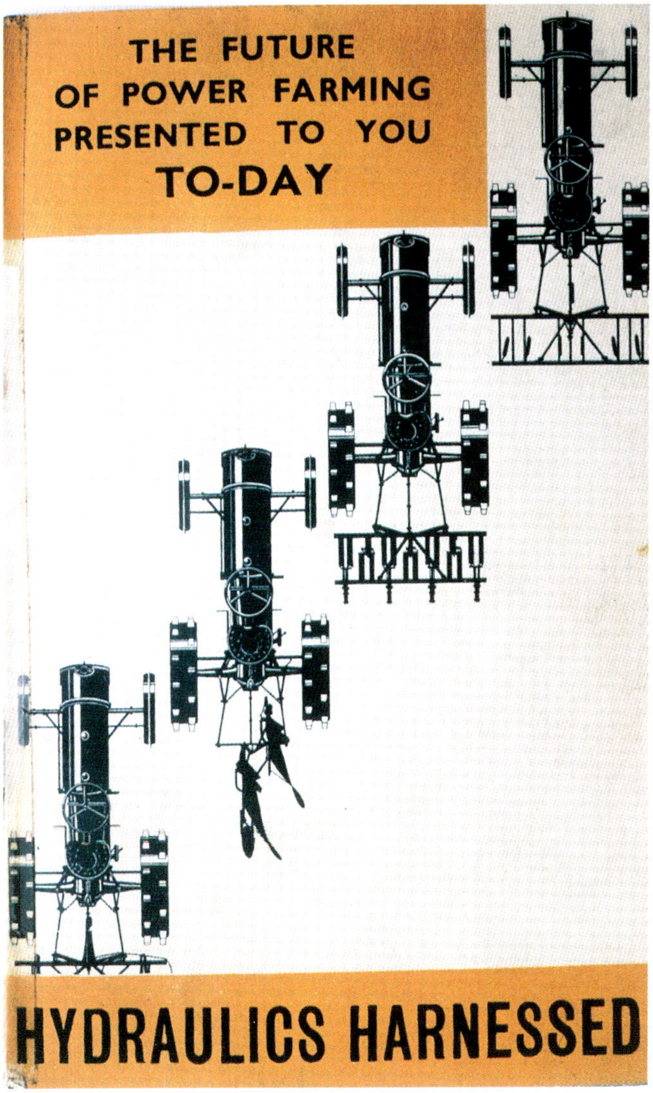

The essence of the Ferguson tractor and Ferguson system – hydraulic control of mounted implements. This very early advertisement shows the first four classic Ferguson implements of cultivator, tiller, plough and ridge – mounted on what appears to be the original Ferguson Black tractor.

Advertising material for the Ferguson Brown tractor was very limited and now commands high prices on the literature market. The first company to produce the tractor was known as Ferguson-Brown Ltd.

Ferguson-Brown Ltd. changed to David Brown Tractors Ltd. within the production run of only some 1,500 Ferguson Brown tractors. The driver has clearly remained relaxed through the company's change of name!

REVISED PRICE LIST—12th JUNE, 1939.

FERGUSON FARM MACHINERY

PRICE LIST

	PRICE	REF.
Tractor on 10″ Steel Spudded Wheels without Hydraulic unit and Linkage (Petrol)	£175 0 0	A
Tractor on 10″ Steel Spudded Wheels with Hydraulic unit and Linkage (Petrol)	£198 0 0	B
Tractor on Adjustable Pneumatics front, 4.5″ × 19″, rear 9″ × 22″, variation of wheel centres from 45″ to 54″, with Hydraulic unit and Linkage (Petrol)	£238 0 0	C
Tractor on Industrial Pneumatics, front 6.00″ × 19″, rear 11.25″ × 24″, including adaptor plates and complete with Hydraulic unit (Petrol)	£246 0 0	D
As above but without Hydraulic unit (Petrol)	£223 0 0	E
Special Paraffin (vaporising oil) Carburetter, if supplied with tractor. Extra … …	£10 0 0	F
Pneumatic Equipment 9″ × 22″ and 4·5″ × 19″ wheel track centres adjustable from 45″ to 54″	£47 0 0	G
Belt Pulley … …	£6 10 0	H
Road Bands … …	£6 10 0	J
Pneumatic Equipment Heavy Industrial Type 11·25″ × 24″ and 6·00″ × 19″ with adaptor plates … … …	£55 0 0	K

DAVID BROWN TRACTORS LTD. HUDDERSFIELD

The Company reserve the right to alter list prices at any time and all goods are invoiced at the prices current on the day of delivery.

DBT 106/1 D3906

The need for accessories, and adaptation of a farm tractor to 'industrial' use was recognized in the Ferguson Brown era.

Ford Ferguson

Harry Ferguson (left) with the first mass produced Ferguson concept tractor – the Ford Ferguson. Ford made the tractor which included Ferguson patent items. Most notable of these was the Ferguson hydraulic system and associated three-point linkage but the adjustable front axle design was also a Ferguson invention.

PLOWS

THE FERGUSON-SHERMAN TWO-BOTTOM PLOW

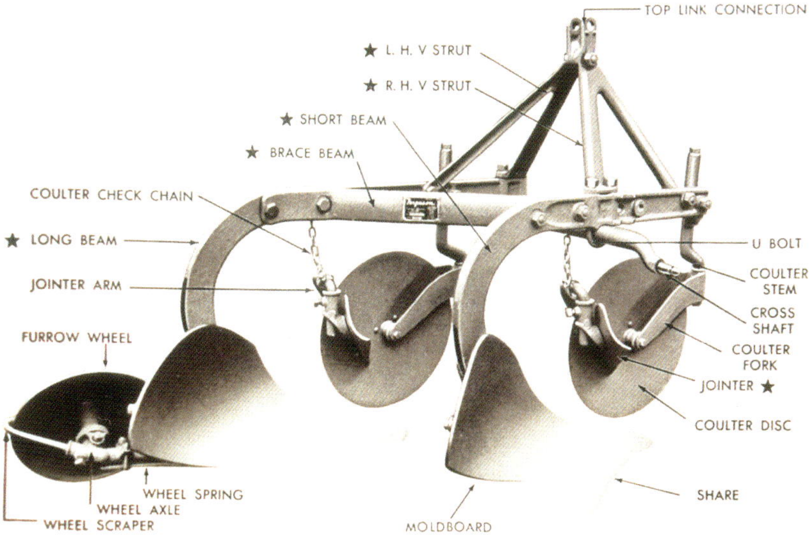

MATERIALS NEW TO FARM IMPLEMENTS PROVIDE LIGHT WEIGHT AND HIGH STRENGTH

• Ferguson-Sherman plows do not have to be built of heavy materials to stay in the soil. The Ford tractor with Ferguson system keeps them there without the use of weight. That is why metals new to farm implements can be used in their construction.

Newly-developed copper silicon alloy steels careful studies of design and highly specialized manufacturing processes have each contributed their important part in the elimination of fuel-consuming, "dead" weight.

By placing added strength where it is needed most and by making other parts lighter, high tensile strength and toughness have been built into these plows with a saving of approximately half the weight formerly considered necessary.

FERGUSON-SHERMAN SINGLE BOTTOM 16" PLOW

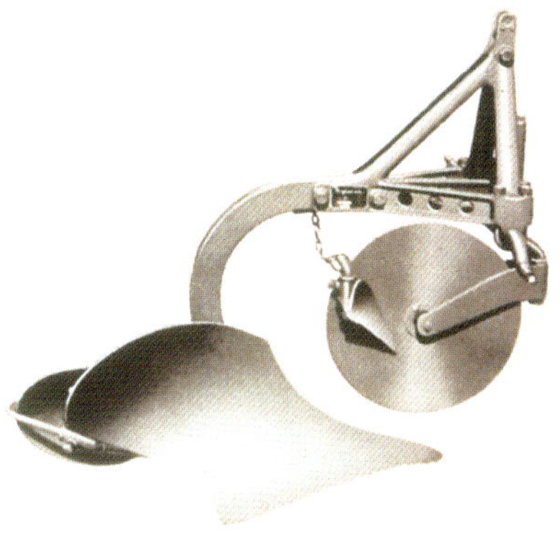

FERGUSON-SHERMAN MFG. CORP., DEARBORN, MICH.

Ferguson's old link with Sherman for plough production continued into the Ford Ferguson era, with the addition of other implements to the range. Ferguson's three-point linkage plough is seen here in its final form – a design which has basically remained unchanged to the present day.

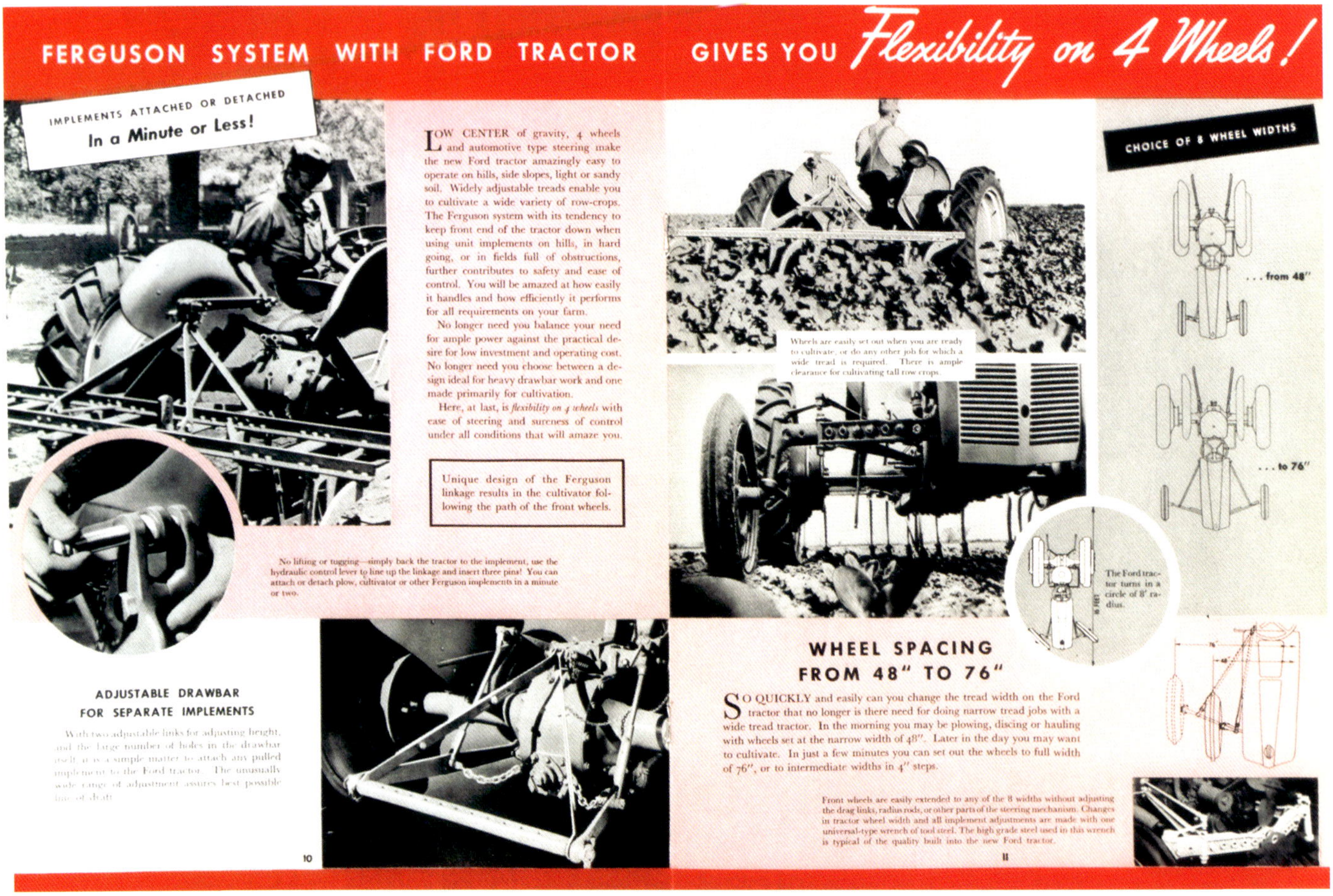

The flexibility of the Ferguson system explained and promoted through photos, text and diagrams.

Sherman also made cultivation implements for the Ford Ferguson tractor. This shot from the 1940s shows the simplicity of the Ferguson linkage and implement concept, and the rowcrop versatility of the tractor.

An idyllic orchard setting for operating a Ford Ferguson tractor and plough combination in the obvious security of a traditional and comfortable family farm environment.

Wood Bros. trailed combine harvesters were an 'approved' machine for use with Ford Ferguson tractors. 1940.

With the Ferguson system the plough was controlled by finger operation of one single small lever at the driver's side. This contrasts with the trailer plough driver's plight where he had to twist backwards through 180 degrees and manually wind or lever relatively large and often stiff control mechanisms. Here, driving a Ford Ferguson is compared to the comfort of the family home.

Only a few implements were offered in the Ferguson Brown era. This changed rapidly with the Ford Fergusons. Four implements shown here make use of the Ferguson hydraulic linkage, and two its rear pto facility. Note that the apparently trailed disc harrows are in fact semi-trailed with an in-built facility for reducing the tractive effort required by lifting the tractor hydraulic arm which reduces the angle of the disc gangs. These discs preceded Ferguson's fully mounted type of disc harrow.

Ferguson in UK

The revolution in tractor design started by Harry Ferguson was to have far reaching implications for rural economies around the world. This 115-page book issued by Harry Ferguson Ltd. in 1952 chronicles many UK farmers' experiences and perceptions of the Ferguson tractor system. The Ferguson tractor is seen here blending in imperceptibly with the family farm in perhaps rural Norfolk in the east of England.

A simple message about the Ferguson tractor. The weight transfer system was a simple concept – but it took the Ferguson genius to identify and apply it! This is a very early British Ferguson brochure showing the early Continental engined tractor and 14 implements.

Many people have argued that the small Ferguson tractor was not the mechanization panacea for all sizes of farm. In retrospect it may be judged that Ferguson should have produced a higher hp version of the TE and TO tractors for large farms, the logic being that other manufacturers successfully sold high hp tractors alongside the Fergusons – sales which Ferguson could have had.

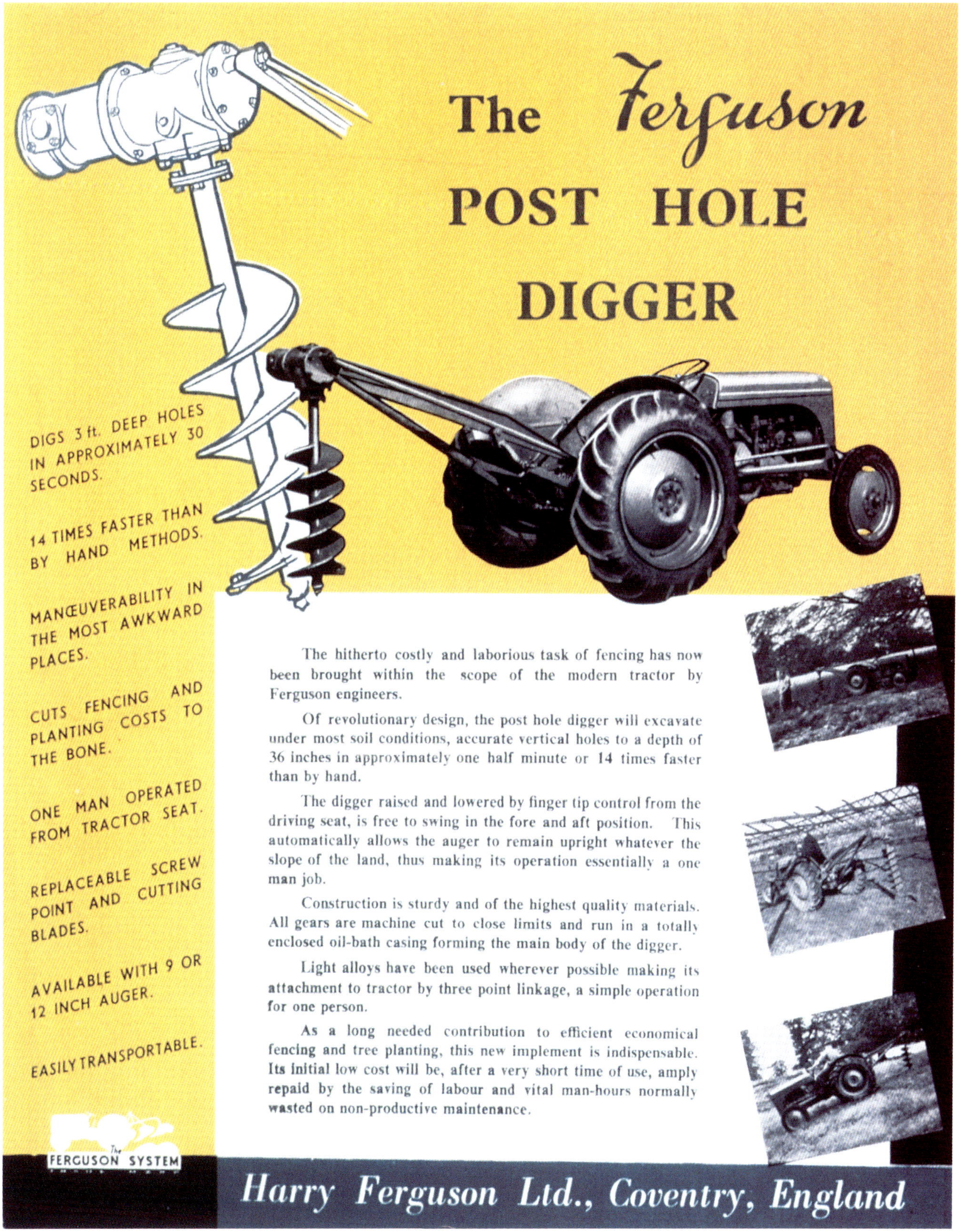

The post hole digger was available with 6-18in. augers. This type of machine was sold well into the MF era.

The earth scoop with its 0.2 cu yd capacity was a popular implement for small earth moving works. MF later marketed a reversible version of the scoop.

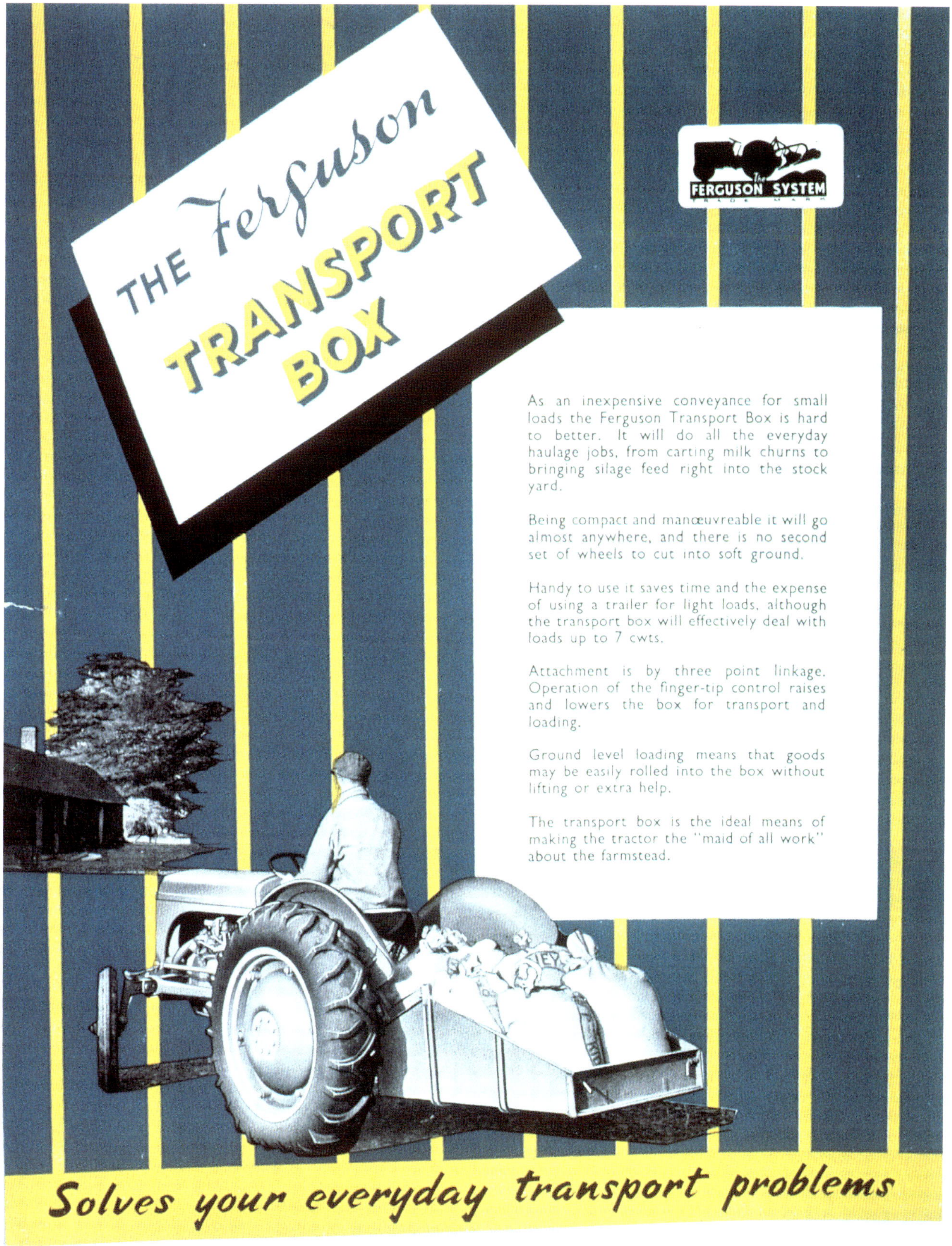

The Ferguson transport box was a boon to small and hill livestock farmers where it was widely adopted. A kit was available to convert the box to a wheelbarrow.

Keeping the home fires burning! These saws, if in good condition, are still much sought after today by smallholders and country dwellers because of their simplicity and mobility.

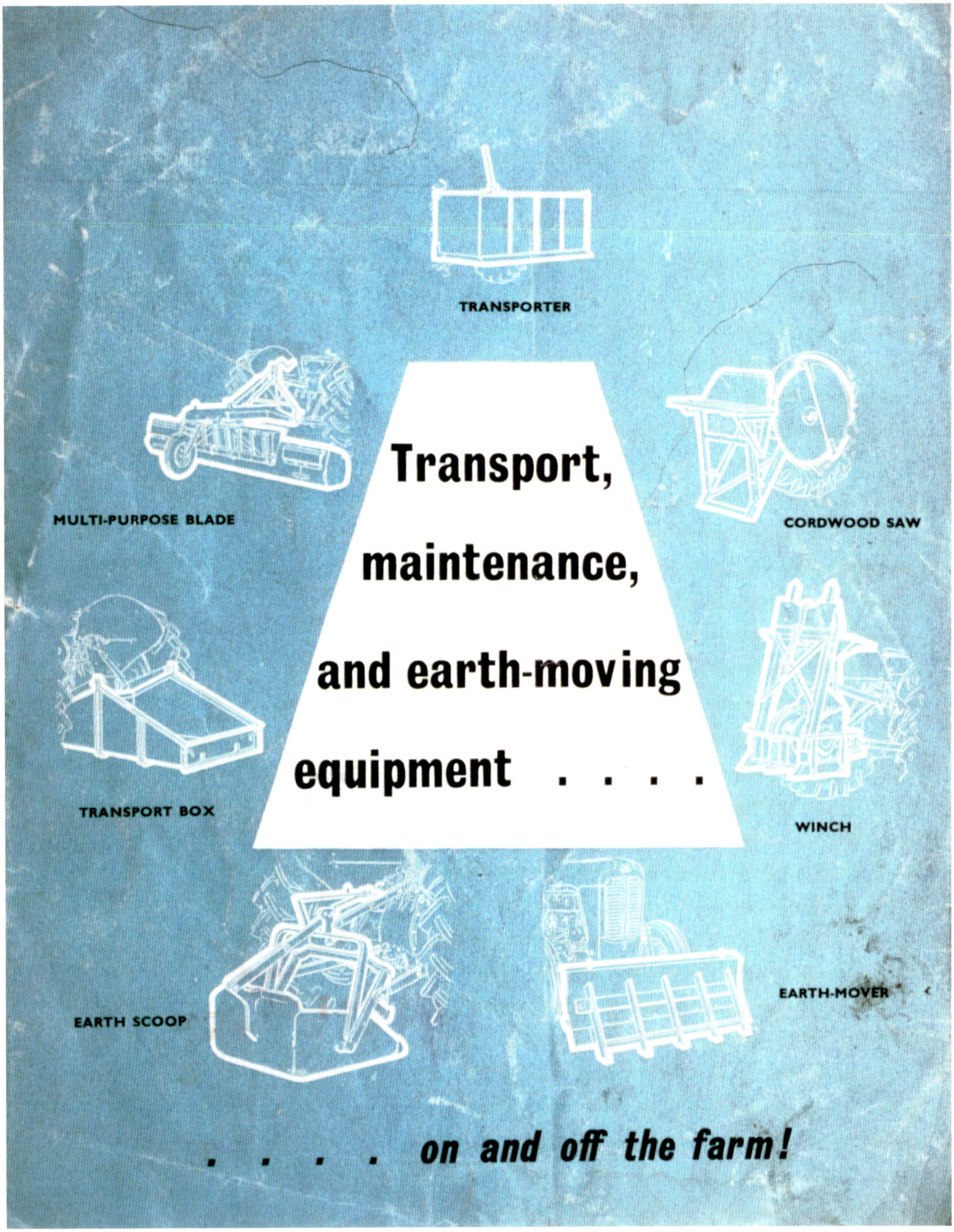

Ferguson was anxious to advertise the capacity of his tractors and implements to perform tasks beyond the general range required for traditional farm field work.

The Ferguson trailer allowed a small tractor to pull the same load on a two wheel tractor that had previously required a larger tractor and four wheel trailer.

The two wheel Ferguson trailer concept was almost as much of a revolution in weight transfer concept as the Ferguson cultivation implements. They have a 3 ton capacity with a high proportion of their weight transferred to the tractor to improve traction.

The L-UE-20 manure loader with a capacity of 600lb was the first to be marketed by Ferguson in the UK. Its design is unusual in that the rams contract to raise the loader. Output per man is stressed here with the combination of loader and Ferguson manure spreader.

The Ferguson high lift loader was very popular for council as well as farm work, and was commonly known as the 'banana loader'. It could be fitted with an 8 tine fork or a 0.33 cu yd bucket.

The Ferguson mower was a most successful and widely adopted implement. It was available as a five or six foot model.

The Ferguson seed drill proved a reliable machine. The drill is hitched to but not mounted on the three-point linkage. The drill is put in or out of work by lowering or raising the lower links.

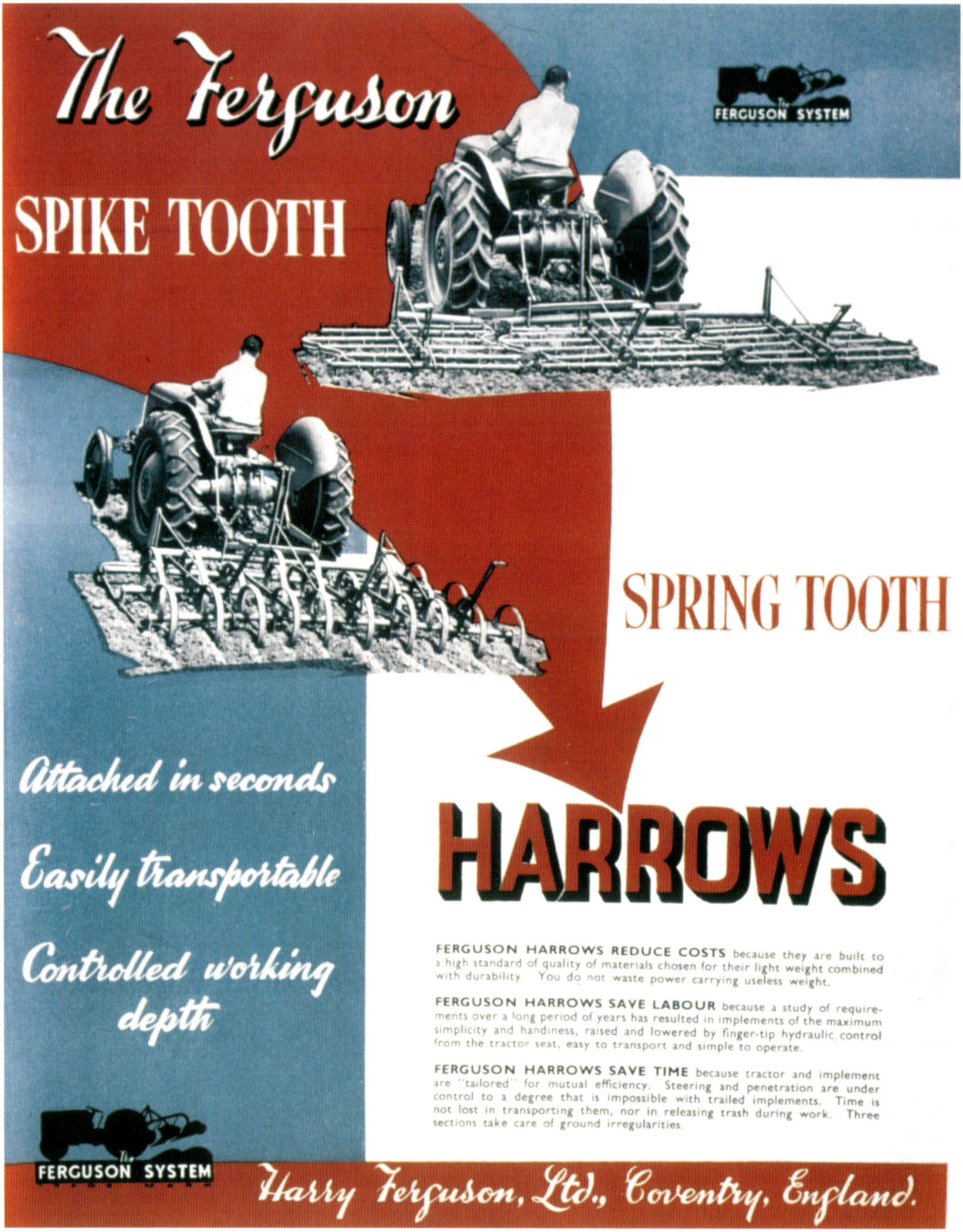

Stressing savings in costs, labour and time by the use of Ferguson system spike tooth or spring tooth harrows for seed bed preparation.

The Ferguson tiller has spring loaded tines and was therefore ideal for stony conditions. A narrower version was made for orchard work where roots can be a problem.

High precision weeding is shown here with this Ferguson steerage hoe. Close weeding was achieved by the hoe operator finely adjusting the line of the weeder as it travelled along the crop rows.

The very effective Ferguson weeder was to become obsolete with the advent of the herbicide in the late 1950s. However, this type of implement has seen something of a revival with the advent of organic farming practices.

The deep digger, single furrow reversible plough was commonly known as the 'Butterfly plough'.

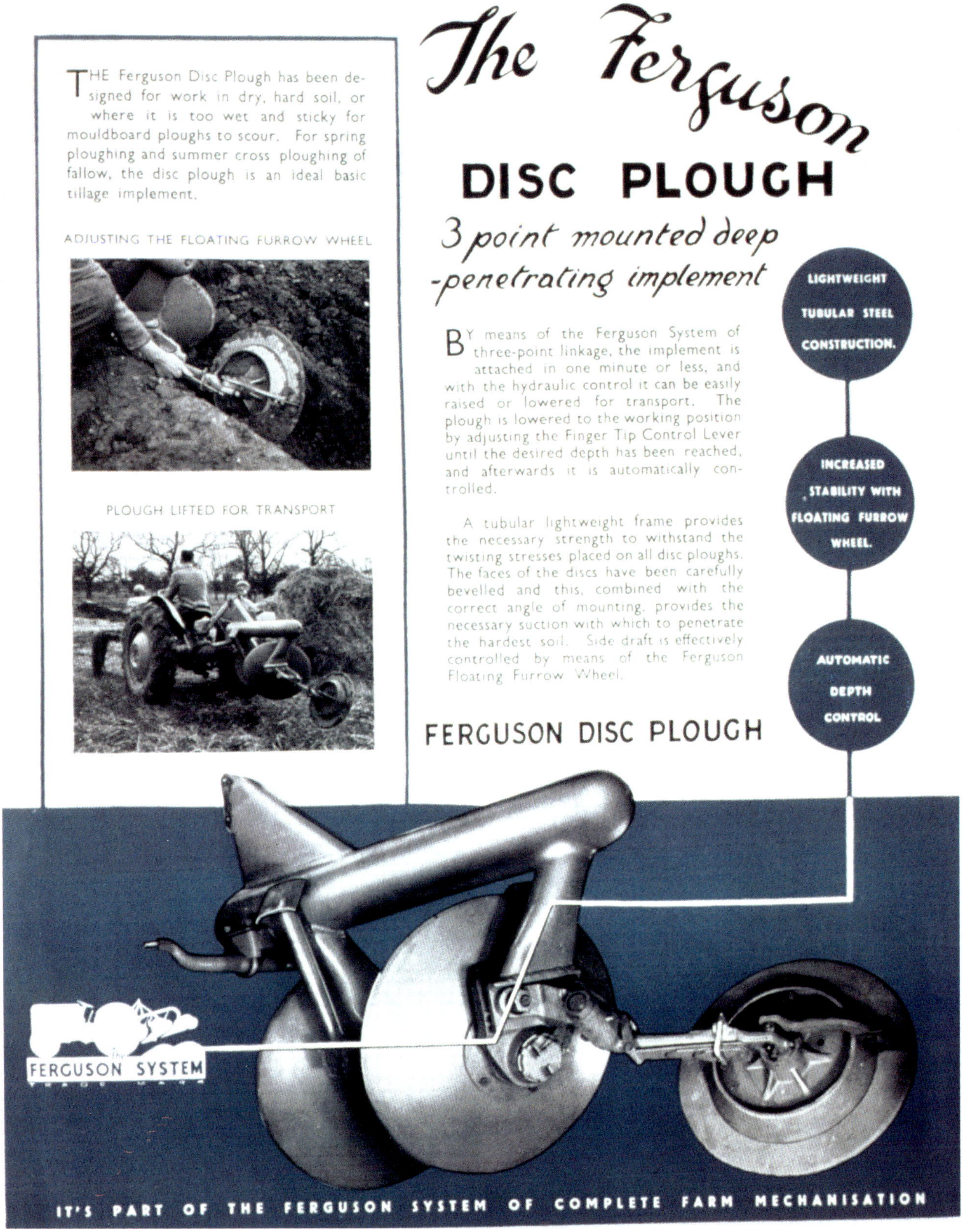

Two and three furrow disc ploughs were made for work in hard dry conditions, on stony or stumpy land. They are also useful in wet sticky land where mouldboard ploughs will not clean.

Perfect ploughing with the Ferguson system in Denmark.

The advantages and features of Ferguson ploughs fully explained! So successful was the Ferguson plough concept that the tractor with plough was adopted as the Ferguson system logo from early days.

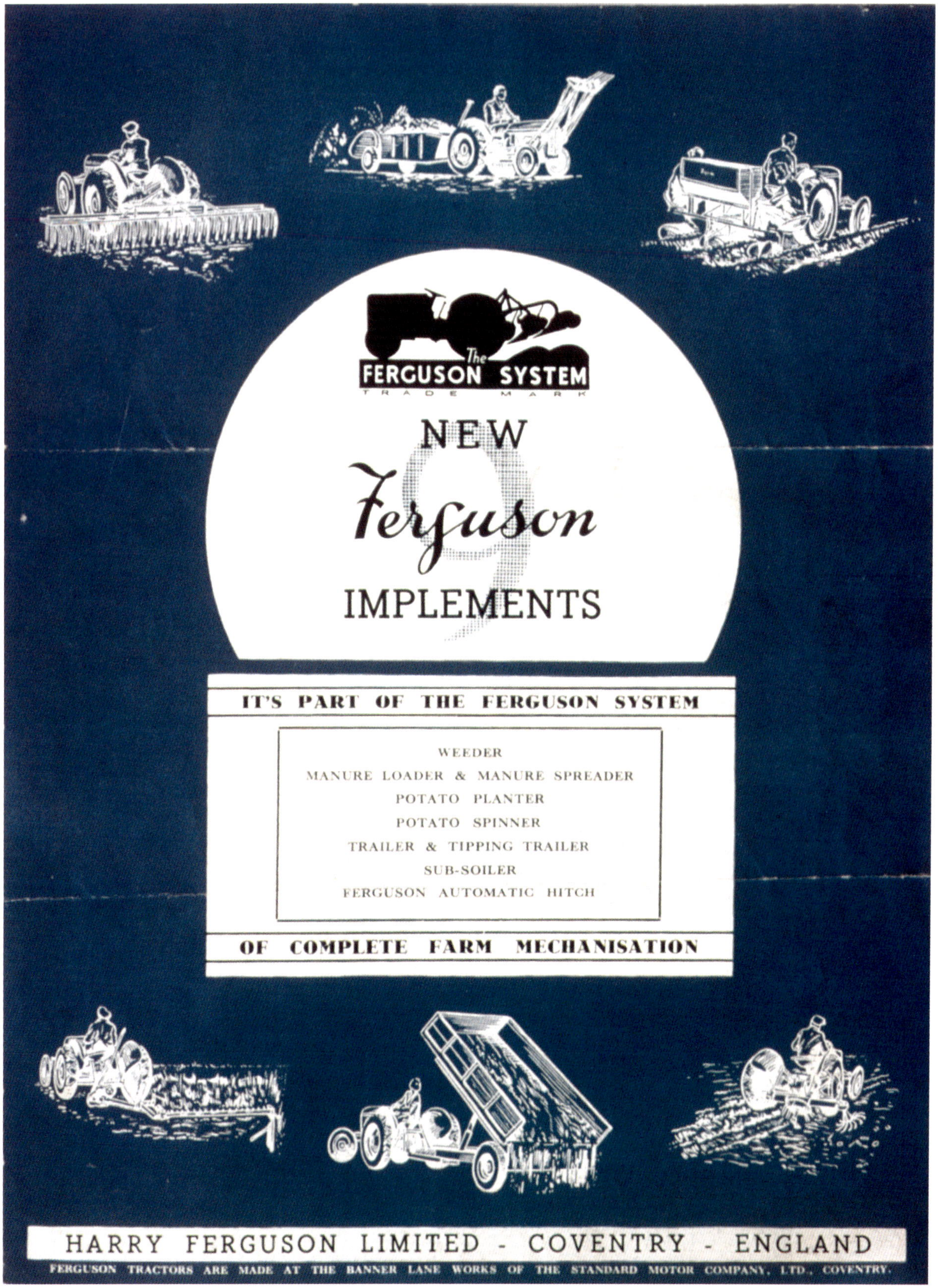

After establishing Ferguson tractor production in the UK, the range of implements was rapidly widened.

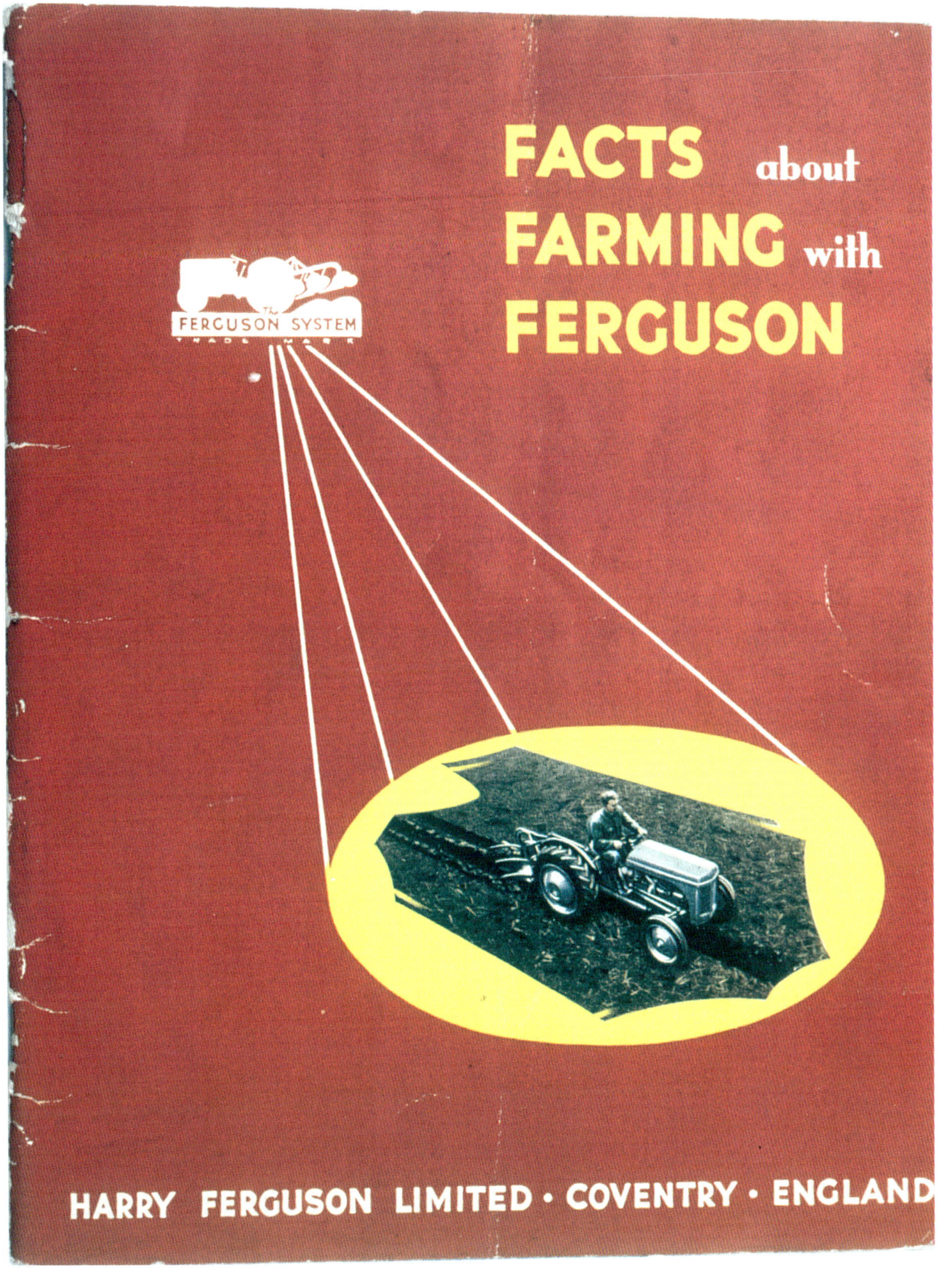

Hard facts for the farmer to digest about the advantages of Ferguson system farming. In general, in the UK Ferguson period, advertising material tended to be more simple, direct and informative than in any other period of MF history.

The Ferguson tractor broke new ground in making tractors a versatile power unit for farmers. Some 70 basic implement types were made for the UK Ferguson tractors!

Advertising the Ferguson tractor and Ferguson system in Denmark.

Ferguson missed no opportunity for designing an implement to meet almost any farm situation. This earth leveller was widely used in semi-arid situations for soil conservation activities. The advertisement comes from 1951.

THE *Ferguson*
SHALLOW COVERING BODIES

THE FERGUSON SYSTEM

GREAT STRENGTH

LONG WEARING QUALITIES

EXCEPTIONAL POWERS OF PENETRATION

EASY ADJUSTMENT

FLAT TOPPED RIDGES

Designed to fit as alternatives to the standard Bodies, these are primarily intended for use when covering Seed Potatoes. When used for this purpose these bodies have advantages over the conventional types. The mouldboards are provided with an easy adjustment ensuring complete and thorough covering of seed even when planting at minimum depth.

Use of the bodies for normal ridging ensures flat type ridges well suited for easy splitting. The soil is lifted rather than pushed to form a looser ridge than is obtained by normal ridging bodies.

From these many important features the farmer will readily appreciate that these are yet another contribution to the fine range of farming equipment of the Ferguson system.

IT'S PART OF THE FERGUSON SYSTEM OF COMPLETE FARM MECHANISATION

Harry Ferguson Ltd., Coventry, England

PRINTED IN ENGLAND FP233/10,000/11/52

Shallow covering bodies were for specialist application and are a rare find today. 1952.

'Industrial' tractors were offered as far back as the Ferguson Brown era. This is a 'full' industrial version of the British TE tractor.

THE FERGUSON INDUSTRIAL TRACTOR

HYDRAULIC BRAKES

HINGED MUDGUARDS

AVAILABLE PETROL, V.O., DIESEL OR LAMP OIL (LAMP OIL ENGINE, EXPORT ONLY) FOR FACTORY, DOCKS, QUARRIES, ETC.

With the introduction of the Ferguson Industrial Tractor the advantages of economy and power of the Ferguson System can now be applied in coal yards, docks, quarries and factories.

Conforming to all the legal requirements for public road work the Ferguson Industrial Tractor is available with either petrol, V.O. or lamp oil engine. (Lamp oil engine for export only).

Stylish mudguards give the driver complete protection from mud at road speed. With the quickly detachable mudguards removed, the tractor is instantly ready for all normal agricultural purposes, making it a truly dual purpose machine and a boon to those who require to do varying types of work.

SPECIFICATION

CONSTRUCTION
As standard with quickly detachable mudguards.

ENGINE
Available petrol, V.O., diesel or lamp oil. (Lamp oil, export only).

GEAR BOX
Constant mesh gear, 4 speeds forward, 1 reverse.

POWER TAKE-OFF
Central shaft

BRAKES
Two indepen

CONTROL
Hydraulic fin

DRAW BAR
Adjustable t
ment.

STARTER
Automobile
operated by

DIMENSIONS.
Height, 54½ i

IT'S PART OF THE FERGUSON SYSTEM OF COMPLETE FAR

HARRY FERGUSON LTD., COVENTRY, EN

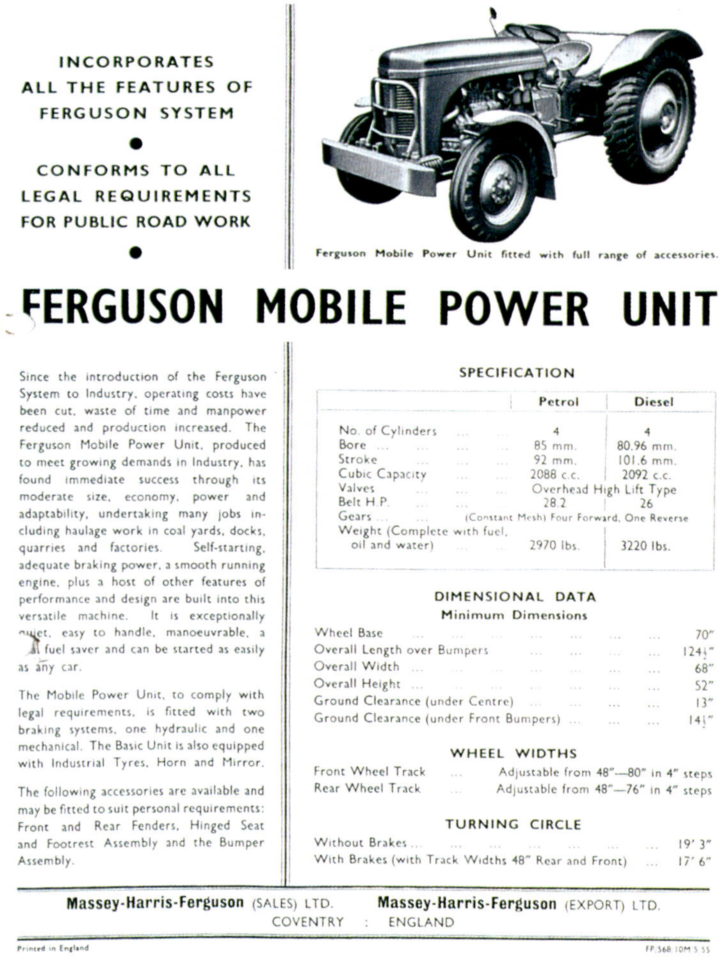

INCORPORATES ALL THE FEATURES OF FERGUSON SYSTEM

•

CONFORMS TO ALL LEGAL REQUIREMENTS FOR PUBLIC ROAD WORK

•

Ferguson Mobile Power Unit fitted with full range of accessories.

FERGUSON MOBILE POWER UNIT

Since the introduction of the Ferguson System to Industry, operating costs have been cut, waste of time and manpower reduced and production increased. The Ferguson Mobile Power Unit, produced to meet growing demands in Industry, has found immediate success through its moderate size, economy, power and adaptability, undertaking many jobs including haulage work in coal yards, docks, quarries and factories. Self-starting, adequate braking power, a smooth running engine, plus a host of other features of performance and design are built into this versatile machine. It is exceptionally quiet, easy to handle, manoeuvrable, a fuel saver and can be started as easily as any car.

The Mobile Power Unit, to comply with legal requirements, is fitted with two braking systems, one hydraulic and one mechanical. The Basic Unit is also equipped with Industrial Tyres, Horn and Mirror.

The following accessories are available and may be fitted to suit personal requirements: Front and Rear Fenders, Hinged Seat and Footrest Assembly and the Bumper Assembly.

SPECIFICATION

	Petrol	Diesel
No. of Cylinders	4	4
Bore	85 mm.	80.96 mm.
Stroke	92 mm.	101.6 mm.
Cubic Capacity	2088 c.c.	2092 c.c.
Valves	Overhead High Lift Type	
Belt H.P.	28.2	26
Gears	(Constant Mesh) Four Forward, One Reverse	
Weight (Complete with fuel, oil and water)	2970 lbs.	3220 lbs.

DIMENSIONAL DATA
Minimum Dimensions

Wheel Base	70"
Overall Length over Bumpers	124½"
Overall Width	68"
Overall Height	52"
Ground Clearance (under Centre)	13"
Ground Clearance (under Front Bumpers)	14½"

WHEEL WIDTHS

Front Wheel Track	Adjustable from 48"—80" in 4" steps
Rear Wheel Track	Adjustable from 48"—76" in 4" steps

TURNING CIRCLE

Without Brakes	19' 3"
With Brakes (with Track Widths 48" Rear and Front)	17' 6"

Massey-Harris-Ferguson (SALES) LTD. Massey-Harris-Ferguson (EXPORT) LTD.
COVENTRY : ENGLAND

Printed in England FP/568/10M/5/55

This 'Mobile power unit' is in fact a 'semi industrial' version of the TE tractor.

This 18-page brochure indicated how Ferguson tractors and implements could be used in building, road making, road maintenance, factories, docks, quarries and dams, airfields, forestry, parks and playing fields, etc.

This single sheet leaflet from 1949 shows the 10in. wide alternative open steel wheels on the rear. The narrow open tyre wheels preceded the availability of narrow pneumatic tyred row-crop wheels.

The Ferguson mounted disc harrow superseded the original semi trailed version. As well as the advantages claimed here, it was also much easier and quicker to move from site to site – the unstated message which is clearly shown here. This 1954 brochure was produced by Massey-Harris-Ferguson.

Twenty-two accessories are illustrated in this brochure. Farmers over the years have progressively demanded an ever increasing range of tractor accessories – for convenience or to diversify a tractorís capabilities. Accessories have now, of course, moved into the electronic era.

The hammer mill was one implement which did not utilize the weight transfer principle. However mounting on the three-point linkage made for easy transport from site to site.

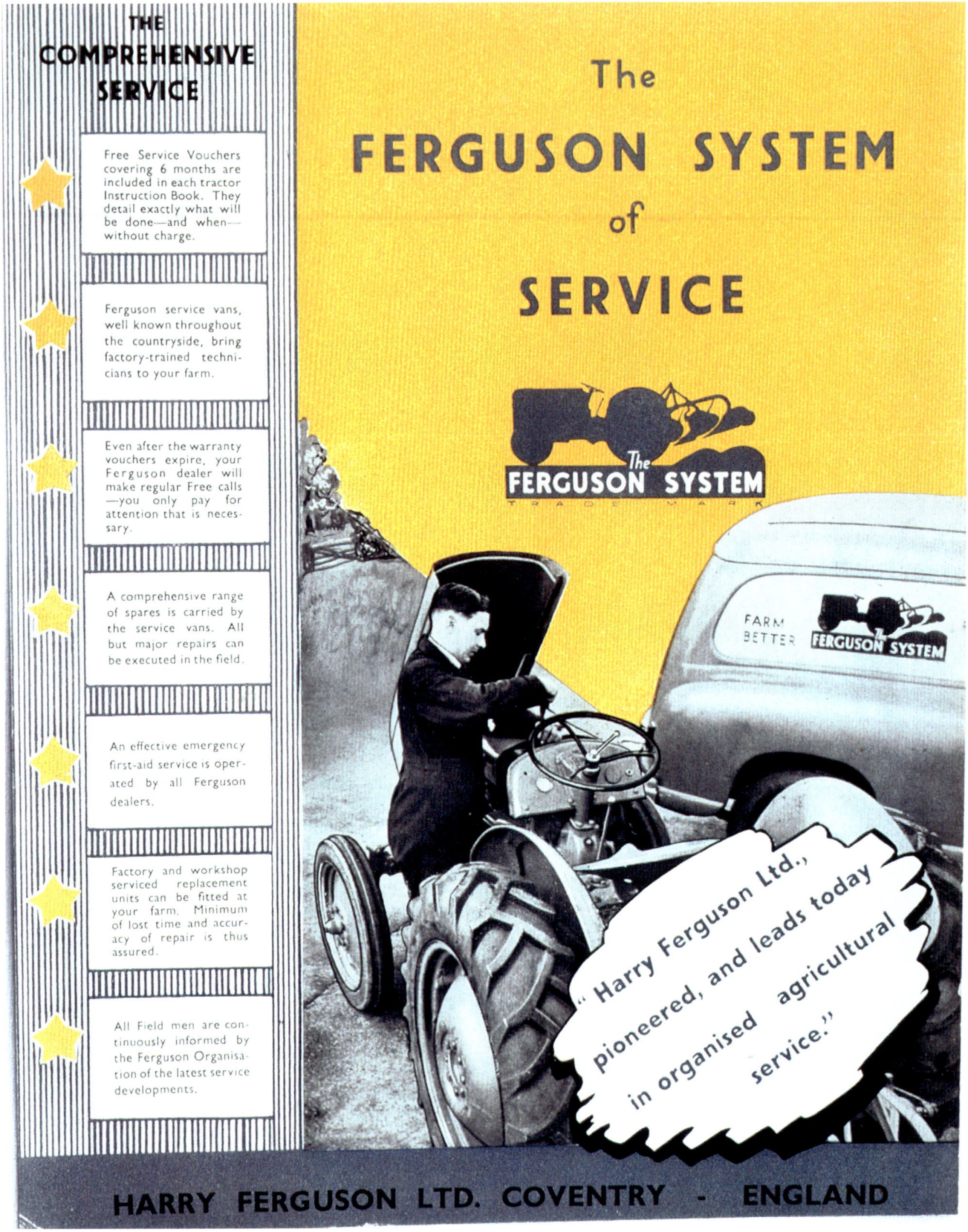

Harry Ferguson placed great emphasis on providing a good after sales service. He is reported on many occasions, when travelling through the countryside, to have stopped and helped or advised farmers with tractor and implement operation.

An advertisement from the Massey-Harris-Ferguson era shows the Ferguson pulling three furrows up a gradient with consummate ease. The brochure listed a range of bases for mouldboard ploughs of 8in., 10in., 12in. and 16in., and two and three disc ploughs.

The Ferguson 35 brought two lever hydraulic control to the Ferguson system – and great happiness to the operator! Another change was the placing of both the independent turning brakes under control of the right foot. Ground speed pto also came in with this tractor.

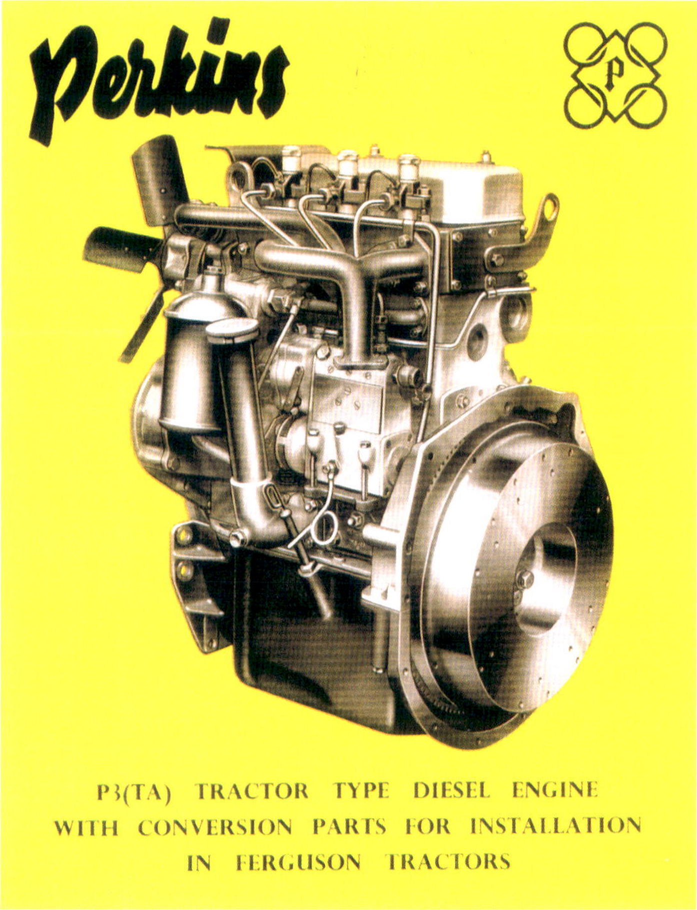

M-H used six cylinder Perkins diesel engines in UK-produced 744 tractors. Ferguson never used Perkins engines, but Perkins very successfully marketed the 32bhp three cylinder diesel engine as a conversion for Ferguson petrol and TVO tractors in the UK. This engine type was adopted for the second version of the MF 35 tractor in the UK and USA, replacing the four cylinder Standard 23C diesel engine or gasoline engines in the USA. MF bought Perkins in 1959. The P3 engine, or A3.152 successor engines as they are known, became one of the world's most widely adopted engines and is still much used today.

Ferguson in the USA

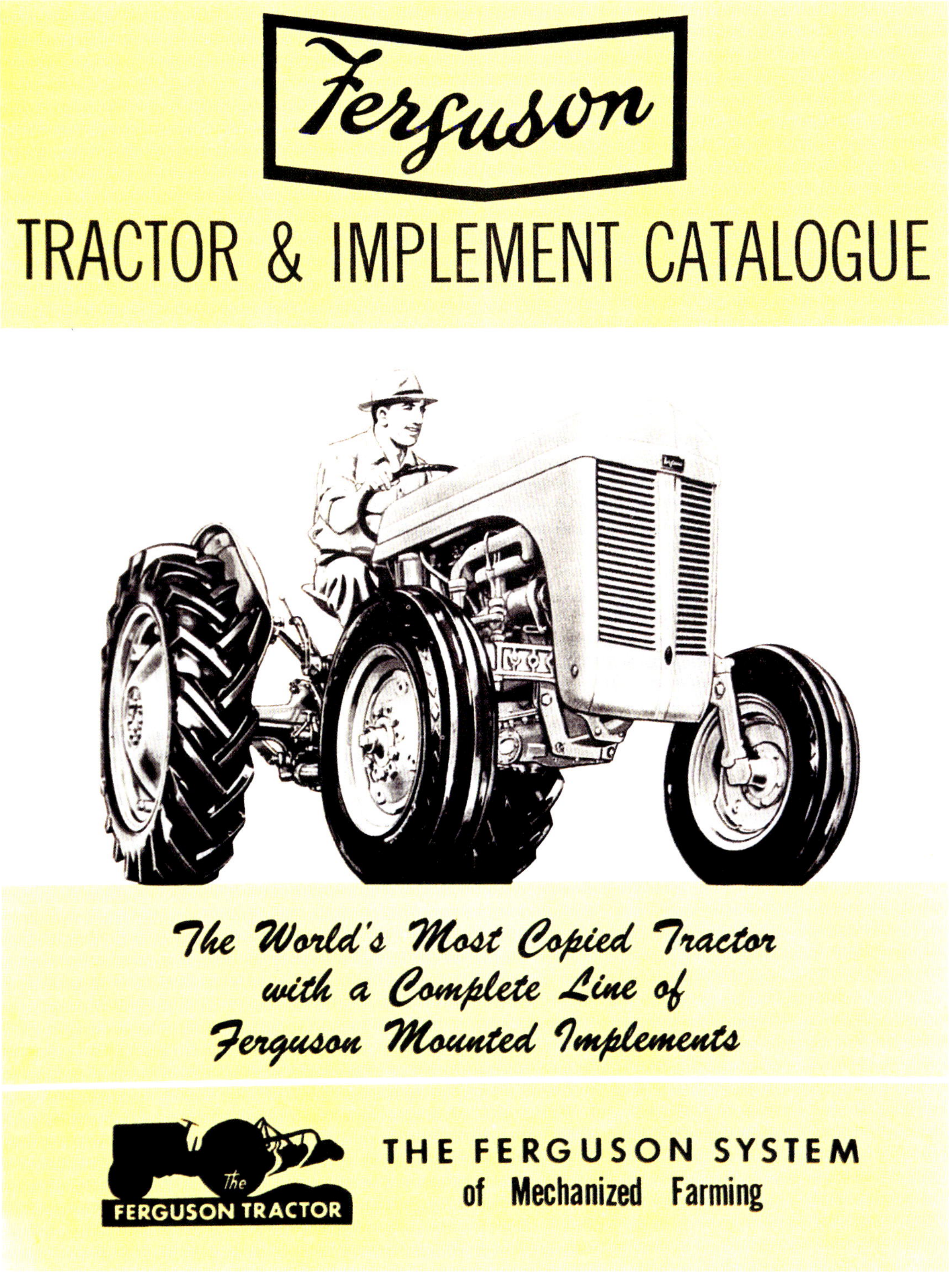

Some UK-made TE Ferguson tractors were sold in North America before Ferguson started production of USA-produced TO Fergusons. This is an early North American advertisement for a British-built Ferguson.

This light duty North American loader is very similar in style to the L-UE-20 loader in the UK, but there are many differences in detail. The Ferguson loader and spreader combination relieved much previous hard work. The tractor looks suspiciously as if it may be an early UK TE Ferguson because there are no side badges on the bonnet. 1949.

When UK-built TE tractors had their piston bore increased from 80mm to 85mm in 1951, the tractor was designated the Ferguson 'Twenty 85' for the North American (mainly Canadian) market.

Happy, mixed enterprise family farming in Pennsylvania with a Detroit-built Ferguson.

Early advertising for a TO Ferguson in North America. Note the now almost unobtainable single rib front tyres which were fitted to early Ferguson tractors in both the USA and the UK.

Both tillers are seen fitted with North American crop guards, which were not marketed in the UK Ferguson implement range.

The rotary hoe could be used for general seed bed preparation or, as shown here, for inter-row cultivation. Very few of these were sold in the UK. 1953.

Newness and power are emphasized for this Ferguson TO 30 tractor introduced in August 1951. The Ferguson badge on the side of the bonnet easily distinguishes American made TO Fergusons from their British made TE cousins. Here the tractor is seen ploughing down a heavy pasture growth in very clean manner.

Ford Ferguson advertising made great play on 'Finger tip control', and the ability of Ferguson tractors to work into very tight corners of fields. These themes were carried on into the TO 30 tractor era. 1952.

Towards the end of the Ferguson era, side-mounted implements were developed for Ferguson tractors in the USA. In the UK this trend only saw prototype development work for a combine harvester. This forage harvester was in a 1954 Massey-Harris-Ferguson brochure.

Only limited sales of the side-mounted baler were achieved. It could also be fitted to Ferguson 40 tractors. As well as a pto drive version, an auxiliary Ferguson engine could be fitted to the baler. 1954.

A rear view of the Ferguson side mounted baler in action, as advertised by Massey-Harris-Ferguson in 1956.

The Ferguson mower was widely adopted on both sides of the Atlantic and was equally successful. 1952.

The trusted Belle City Corn Picker was sold as a Ferguson implement, but retained the Belle City badging. In this 1950 picture the machine is seen working on a cold autumn morning.

The side delivery rake was particularly popular with alfalfa growers in the USA because of its gentle handling of the crop, but is here seen turning grass hay in a Massey-Harris-Ferguson era brochure. 1956.

The flexi tiller seen operating here certainly lives up to the claims by producing a tilth from large discs operating to plough depth. Some were fitted with seeder boxes for an all in one operation. 1955.

The drill planter, here shown with fertilizer applicators, was available with a wide range of openers and attachments, including a peanut planting attachment.

The Ferguson tiller was widely used for general cultivations rather than inter-row work. Its particular features were the spring loaded tines which would lever back out of work when they hit a large stone or tree root, thereby avoiding damage. It is shown here working in 1957, the year in which Massey-Harris-Ferguson changed its name to Massey-Ferguson.

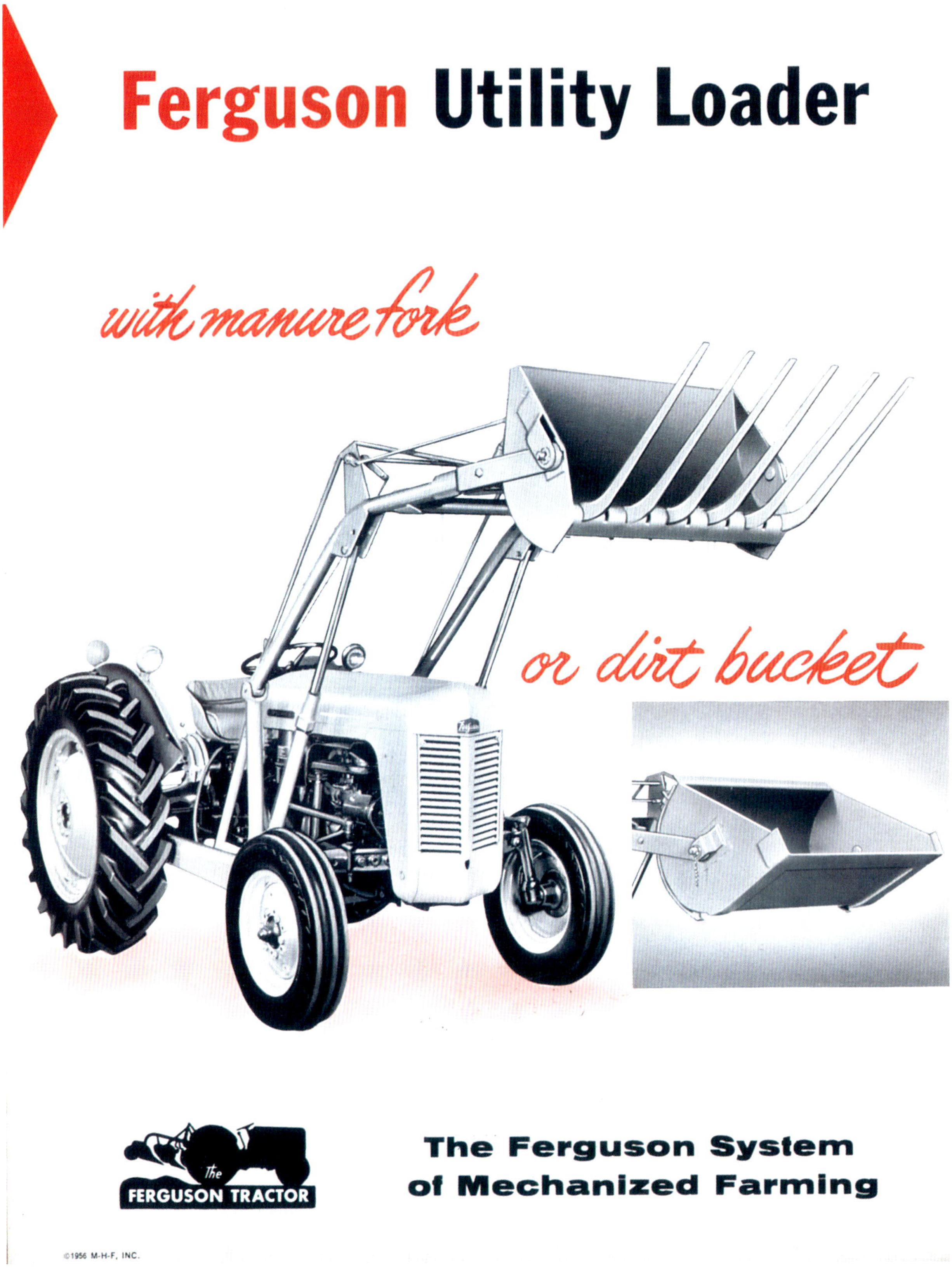

The utility loader offered in North America was similar to the 735 loader offered in the UK for the Ferguson and MF 35 tractors, but the curved tines are distinctly North American. 1956.

Making no use at all of the Ferguson weight transfer system! This two implement rig of a trailed Belle City Corn Picker and trailed Ferguson four wheel trailer would never have been seen in the UK where the Ferguson two wheel trailer was widely adopted. 1952.

No end to the versatility of the Ferguson tractor! Bombardier half tracks gave the normally wheeled tractor crawler capability in such situations as snow, heavy earthmoving or difficult hillsides – all where added traction was vital. Full track versions enabled British-made TE Fergusons to cross Antarctica in 1957/8, and later US-made MF 65 tractors to undertake polar work. 1951.

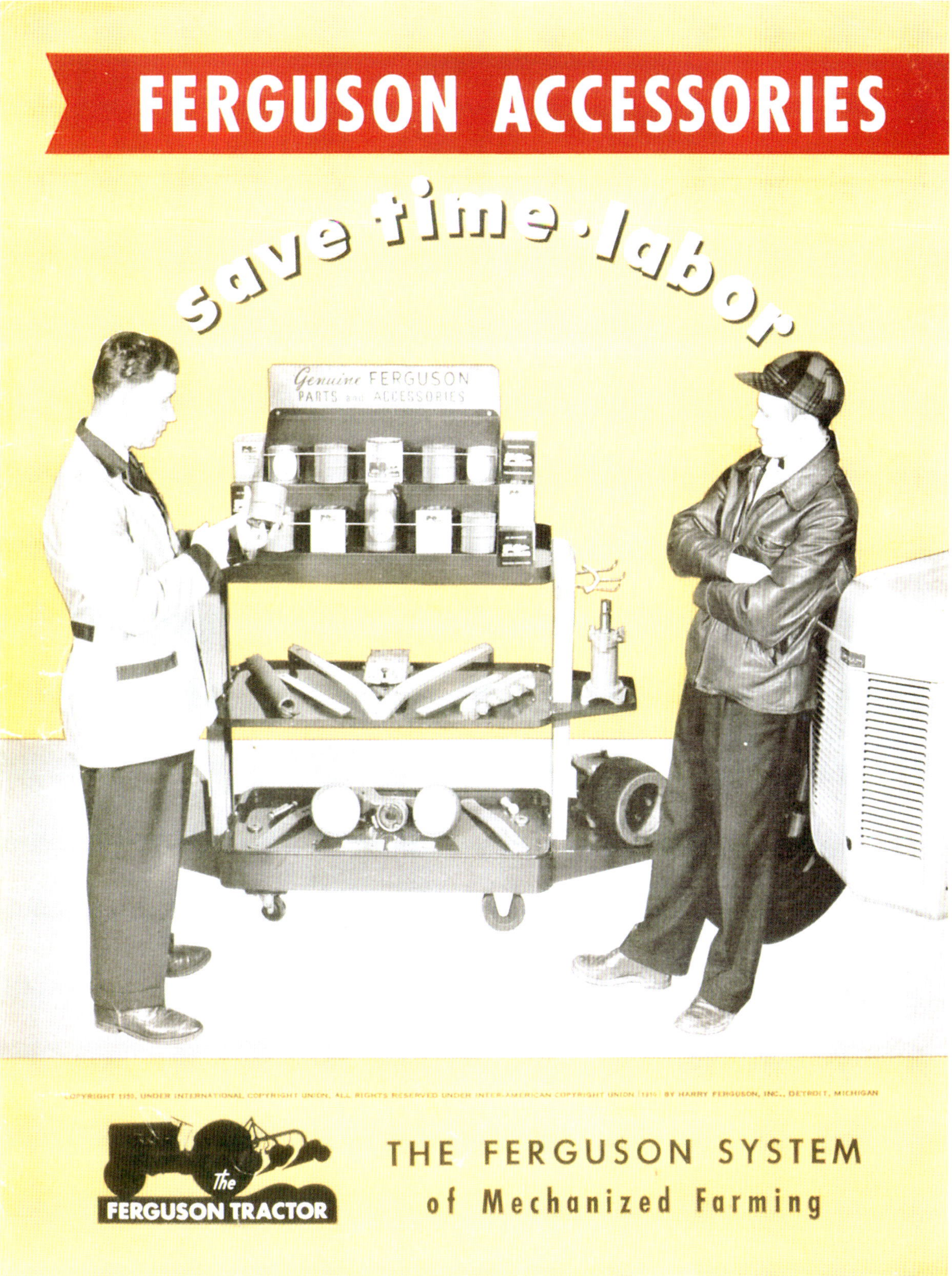

Ferguson offered a wide range of accessories in both the UK and the USA. One implement accessory which was possibly offered uniquely to the North American market was a rolling fin assembly for cultivators and ridging ploughs. A Ferguson badged battery (in the case moulding) was also offered. 1950.

The Ferguson 40 tractor was the last Ferguson badged tractor to be produced in the USA. Only 9,000 or so were made and it is now quite a rare collector's item. During the two line M-H-F era it was sold in red Massey Harris livery as the M-H 50, then subsequently it evolved to become the MF 50. The Ferguson 40 broke with Ferguson tradition by being made in standard, high clearance and tricycle form. 1956.

A Ferguson 40 pulling the trailed Ferguson multi-flow drill. Earlier versions of this drill in the UK were semi trailed by a linkage which enabled the tractor hydraulics to lower and raise the coulters in and out of work. 1956.

Superb row crop vision with a mid-mounted cultivator on a tricycle type of Ferguson 40 tractor. 1956.

A standard version of the Ferguson 40 tractor pulling the little known Ferguson F12 baler in the M-H-F era. Instead of using the pto drive, the baler could be fitted with a two cylinder air cooled, 14hp Wisconsin engine. 1957.

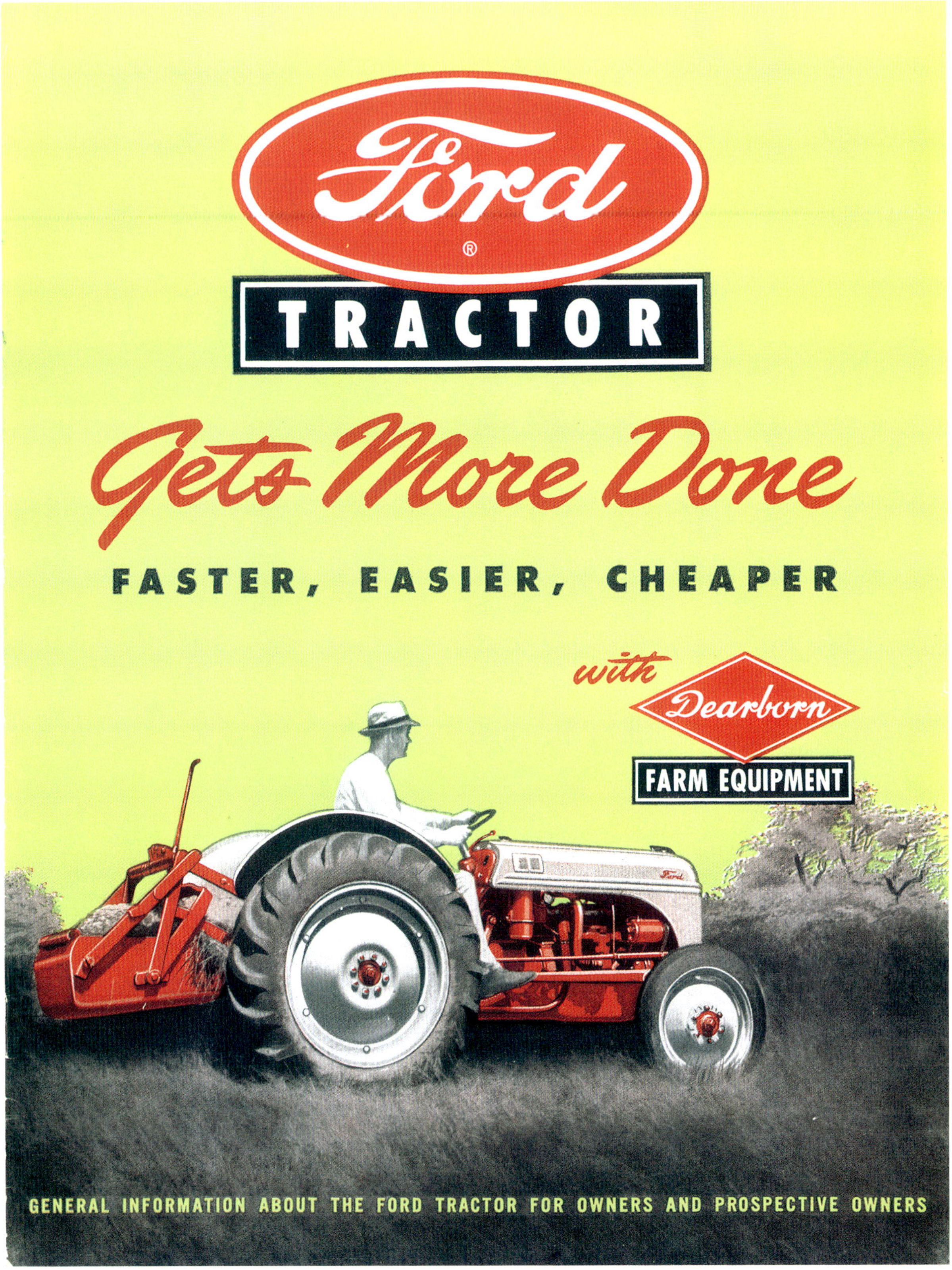

The Ford 8N tractor – undoubtedly a Ferguson looking machine! Production of this tractor resulted in Ferguson successfully suing Ford for infringements of his patents. The tractor's resemblance to the old Ford-Ferguson tractor is quite apparent. 1951.

Section V

Massey-Harris-Ferguson 1953–1957, Massey Ferguson 1957–1994 and AGCO 1994–

(holders of the Massey Ferguson brand name and the current producers of Massey Ferguson equipment)

Massey-Harris and Ferguson merged in 1953 to become Massey-Harris-Ferguson. The merger added a much needed, unique and dominant tractor and implement design to the M-H portfolio, which had become weak in the tractor sector. It was a superb complement to M-H's strengths in combine harvester design and manufacture at the time.

Sadly, the Harris name was dropped in 1957 and the company was simply titled Massey-Ferguson. The hyphen was later dropped and the name Massey Ferguson continues today. Taken over by AGCO in 1994, there are no longer any Massey Ferguson companies, but AGCO who own the brand name, continue to badge equipment which they make or market as Massey Ferguson products. The AGCO acquisition placed Massey Ferguson in an AGCO portfolio of several major agricultural equipment manufacturers, of which MF products are the major contributor to turnover.

Massey Ferguson continued with the pattern set by its predecessor companies – production of its own equipment and strategic acquisition or marketing of other companies or their products. In this respect some of the more significant relationships have been with Landini (tractors), Mid Western Industries (industrial equipment), Perkins (diesel engines), Eicher (tractors), Hanomag (industrial equipment), Droningborg and Claas (combine harvesters), Gardners (diesel engines), Greenland (grass and planting equipment) and Sambron (telescopic handlers). Some of these relationships survive to the present day, albeit often in a way different from the original.

Massey Ferguson also went on to establish more of their own products, and some licensed manufacturing operations around the world, in both the agricultural and industrial equipment sectors.

1953

A new logo for a new company with global interests. Massey-Harris-Ferguson was to remain the company name for four years before changing to Massey-Ferguson, and later Massey Ferguson.

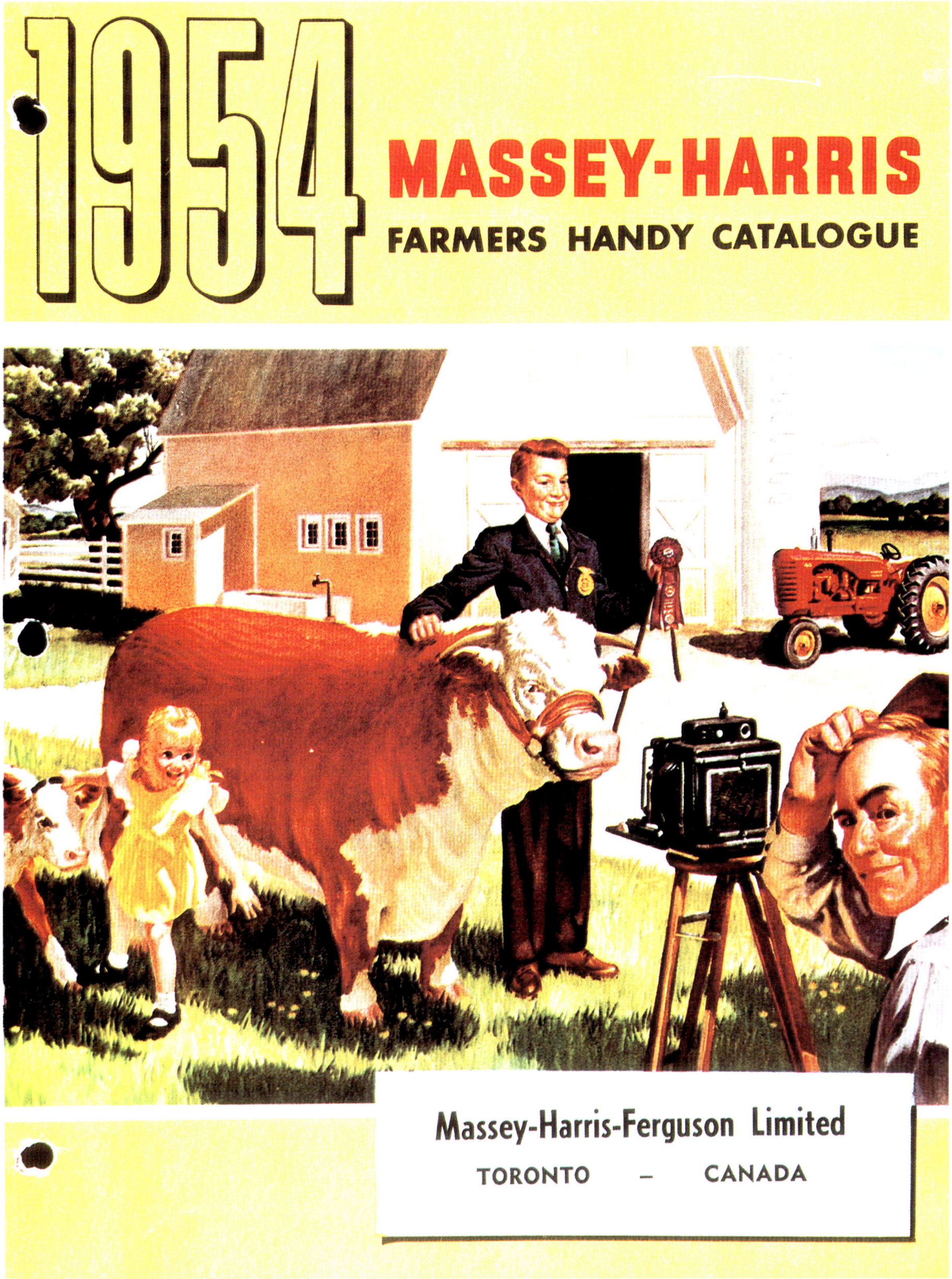

Initially the newly formed Massey-Harris-Ferguson maintained a two line policy with distinct M-H and Ferguson lines, as illustrated by the 1954 M-F *Farmers Handy Catalogue*. Promotion of M-H in the family farm setting continued with a comparison perhaps being drawn between the prize Hereford beast and the powerful M-H tractor.

The 1955 *Farmers Handy Catalogue* is the first after the merger. It conveys the two line image with the separate logos surmounting a country scene.

Massey Ferguson tractors

The MF 35 and 65 tractors – both products of Ferguson design – founded the MF tractor line in the UK. They are shown here with an M-H design combine rebadged Massey Ferguson, and an early MF baler. Perhaps steady evolution from old products to new is the message of this very early UK M-H-F brochure.

Bringing good cheer to the British farmer is the four cylinder Standard Motor Co. diesel engined version of the MF 35 tractor. This tractor grew out of the Ferguson stable, was one of the best selling MF tractors ever, and derivatives of it are still made today under licence in countries such as India.

The MF 35 three cylinder diesel tractor, introduced in the UK in 1959, was not a hard sell for dealers. The Perkins three cylinder A3.152 engine must be one of the most successful diesel engines ever produced, and versions of it are still widely used today.

The MF 35X was the last version of the MF 35 to be produced in the UK. The X indicated that the three cylinder Perkins diesel engine had an improved power rating over the previous MF 35 model. However, in France the original four cylinder MF 35 tractor was marketed as a 35X, the X indicating that it was fitted with a differential lock.

The introduction of multi-power transmission brought on-the-move gear change to MF tractors. Switching between low and high in a given gear gave about a 30 per cent speed increase by convenient finger tip control – literally a 'flick' of a lever, as highlighted here.

The USA-built MF 50 tractor represents MF's final abandonment of a two line M-H and Ferguson policy. Originally designed as the Ferguson 40 it was produced in M-H livery as the M-H 50 and finally became the MF 50 powered by a Continental gasoline engine. 1958.

1958 Massey-Ferguson Guide

TO MODERN FARMING

Introducing the all-new Massey-Ferguson 65

4-PLOW TRACTOR WITH EXCLUSIVE FERGUSON SYSTEM FEATURES

Here's the all-new Massey-Ferguson 65! See more information and pictures inside—along with two other powerful new tractors: the Ferguson 35 and the Massey-Ferguson 50! Now there are Ferguson System tractors in sizes for every farm. The best features of Massey-Harris and Ferguson engineering have been combined to bring you more economical, more maneuverable tractors—all with the famous 4-way work control. There's new Massey-Ferguson equipment too—designed to make your job easier for years to come. See the new tractors and implements now—ask your dealer for more information on the complete Massey-Ferguson line of profit-making equipment.

The MF 50 was uprated to become the MF 65 tractor. This was produced with slightly different styling in both UK and USA. Shown here is a USA-built MF 65 with Continental gasoline engine.

MF 65 tractors were only produced with diesel engines in the UK, but in the USA diesel engines were not adopted until two years after its introduction. This USA model boasts a Perkins AD4.203 four cylinder diesel engine and differential lock. The lock was introduced by MF in the 35 and 65 tractor era.

In France MF emphasized the bond between man and tractor, with MF able to solve all problems with its machines and implements!

Continental Europe has always been a large market for tractors for work in orchards and vineyards. The French-built MF 825 (8 signifies built in France, a carry over from M-H days) was one such example and was powered by a Perkins 4A.107 diesel engine. The minimum tread is 0.71m front and rear and the tractor was produced between 1963 and 1965.

Requirements for high power tractors have surged relentlessly from the late 1950s onwards. In North America MF produced the '5 plow' power (traditional M-H tractor power rating) MF 85 with a 223 cu in. Continental gasoline engine in response to this demand, and claimed it to be the most powerful Ferguson System tractor.

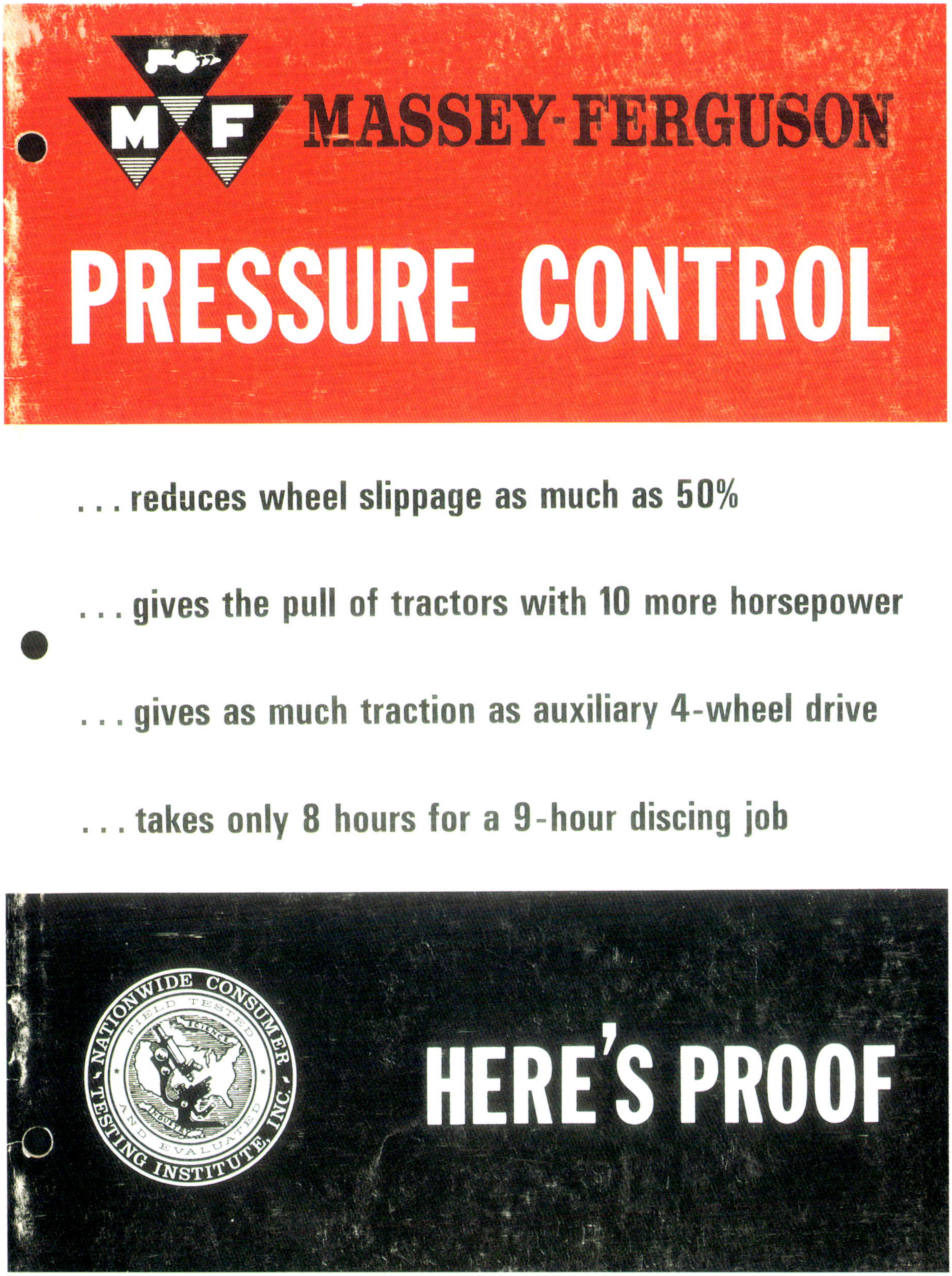

The multipull hitch was perhaps the last significant development of Ferguson's concept for transferring weight from implement to tractor to aid traction. Multipull hitches, of which there were several versions, enabled controlled weight transfer from trailed implements or four wheel tractors. Surprisingly, they were never widely adopted. 1961.

The multipull hitch was operated via the pressure control component of the tractor's hydraulic system. Applying variable lift pressure to the multipull hitch enabled the transfer of a variable amount of weight from implement to tractor. MF clearly spells out the facts in this 1966 advertisement.

MASSEY-FERGUSON

FARMER'S HANDY CATALOGUE

1965—THE YEAR OF THE RUGGED NEW MF TRACTOR BREED

1965 saw the launch of the MF 100 series of tractors worldwide. In Canada the announcement was made in the *Farmers Handy Catalogue*. Here we see the 'Red Giants' (as they were sometimes called) demonstrating their cultivating and seeding abilities on a rich, red soil.

The MF 165 tractor was a huge international success for MF. A UK-produced machine as sold on the South African market is shown here.

MF responded to French small farmers' needs by manufacturing the 130 tractor with a Perkins A4.107 engine which produced 27 pto hp. The tractor was made in France but marketed overseas as well. The advertisement appeared in 1970.

MF's purchase of the Italian Landini company, Italyís second largest tractor manufacturer, gave them access to a crawler tractor range. The 244 shown here has classic Landini styling but has been rebadged MF. It had a Perkins A3.152 or B3.152 three cylinder, 40hp diesel engine and was made in large, standard, narrow and vineyard versions.

One of the second generation of crawler tractors to come from the Landini stable was this 154C with Perkins AD3.152 34.6 kW (max) engine. The brochure, like most South African MF literature, is bilingual – English and Afrikaans.

The MF 1000 series of tractors grew out of the 100 series and adopted their styling. They were produced in response to demand for higher horse power tractors – particularly in North America. The 1150 was MF's first V8 engined (Perkins V8.510 diesel) tractor and was rated at 135hp. An 1155, rated at 14hp, was also made. The 1150 portrays solid, heavy lugging power weighing in at some 11,000lb.

Western style tractors for Prairie and large farm work date back to M-H days. MF followed on with the tradition, producing platform tractors with wide tyres, heavy duty front axles, cast wheel centres and multi-power. They were described as 'a rugged breed – the brawniest, brainiest tractors MF ever built'. 1968.

Stagelike presentation of the 240 tractor by FEDMECH – MF's licensee in South Africa. The 240 was the successor to the 135 tractor still using the Perkins AD3.152 engine. 1981.

LIBYAN TRACTOR COMPANY

Models produced
Aljadah 240 — 47hp
Aljadah 275 — 70hp
Aljadah 290 — 79hp
Aljadah 1010 — 16hp

Plant and facilities
- Located at Tajoura, near Tripoli
- Site area - 21 hectares
- Covered area - 11,000m^2
- Staff (Jan 1986) - 286
- Capacity: 3000 to 5000 tractors per year

Sometimes tractors have to change colours for certain markets! 1986.

One of the larger MF agricultural crawlers to come out of Italian Landini after acquisition by MF was the 1144C. It is fitted with a Perkins 110hp T6.354.1 engine and weighs 7.5t. 1981.

The 234C crawler which was produced in narrow (1.15m), standard (1.32m) and wide (1.48m) versions still used the Perkins AD3.152 engine. The narrow tractor was particularly suitable for orchard and vineyard operations in mainland Europe. 1983.

The 300 tractor series was introduced in the mid-1980s with 'sub series' within the main series. The 307 series, typified by this 397 Australian market version, included the 377, 387 and 397 models. All were four wheel drive, but the 377 was available in two wheel drive form. The 397 used a Perkins A4.248 engine, the others the A4.236.

MF produced four wide rowcrop models of the 300 series tractors from 73–95hp. They could straddle two 48in. rows, achieved by use of deep dish rear rims and optional spacers. 1994.

The MF 300 tractors included a 46 model Italian-made speciality range. There were vineyard (V) models; narrow, short and low (GE) models, narrow and short (S) models, and high clearance (H) models. They were variously designed for vineyard and orchard farmers, and work in bedded plants, vegetables and muddy conditions This catalogue for vineyard tractors illustrates the confined conditions in which they could work.

The MF 300 series of Low Profile and Mudder tractors contains eight low profile and one mudder tractor. The low profile are for orchard work, and the mudder has a special application for irrigated wet soil conditions where high clearance is needed. 1996.

As far as can be ascertained the MF 465 is the only '400' series tractor and was assembled in Holland. It has a Perkins A4.236 engine.

Both three and four cylinder engines were used in the MF 500 series tractor range. This Perkins A4.318 four wheel drive version appears to be making light work of cross slope chisel ploughing in a dry British autumn.

Massey Ferguson

699

Smooth, 6-cylinder power

600 series tractors were made in the UK and France. One of the larger models was this 699 tractor which was available in two or four wheel drive form. 1984.

MF 2000 series tractors were made in France. Shown here are three of the models distributed by FEDMECH in South Africa in one of their English/Afrikaans bilingual brochures.

A French brochure from 1992 for French-built tractors. Eight models are listed in two or four wheel drive form with a weight range of 3565–5630kg.

MF's first on-the-move gear change came with multi-power in the 1960s. This has now advanced to the present day Dynashift which on these 3000 tractors gives four speed changes in each of eight synchronized gears to produce 32 forward and 32 reverse gears when combined with the reverse shuttle. Optional creeper gearboxes enable 48 forward and 48 reverse gears. 1992.

Perkins' new 1000 series engines claim, amongst other features, 12.6 per cent increase in torque, 77 per cent increase in torque back-up and more power at all revs along the power curve. The engines have a four lobed combustion bows – 'Quadram' – to optimize and control air motion in the cylinder. 1993.

The USA-made 4000 series articulated tractors with four equal sized drive wheels were launched in 1979. The range came to include models of over 300hp. They were often fitted with dual wheels for the heaviest of work. Tyre width could be up to 30.5in.

The UK-produced MF 4200 tractor series was launched in 1997 – MF's 150th anniversary. Unusually for the late 20th century black and white advertising has been used. Eight models are listed here in two or four wheel drive form and range from 55 to 99 pto hp. They are the first major new tractor range to be introduced under AGCO's ownership of MF.

The new MF 4200 series includes this low profile, sloping bonnet version. Similar 'high visibility' tractors have been produced in the French-made 3000 'Visioporteur' models 1997.

Both the 6100 and 8100 series tractors are made in France. They have Dynashift or Speedshift transmission, electro/hydraulic lock, are available as two or four wheel drive and have automatic function controls and Datatronic performance monitoring. 1997.

The 10t MF 9240 245hp tractor was the top of the MF tractor range in 1996. It is in fact a tractor produced by the White Tractor Manufacturing Co. in the USA; White is also owned by AGCO.

The 1999 AGCO product guide for Brazil advertises 16 MF agricultural tractors, four MF combine harvesters, two Ideal combine harvesters, the MF 86HS industrial backhoe loader, the Maxion 750 tractor backhoe loader and a lawn tractor. Interestingly, AGCO Brazil appears to be the only operation still producing MF Industrial badged equipment.

Compact and garden tractors

The first turf tractor by MF was based on an early MF 35 tractor of the Ferguson TO 35 tractor type. Rising affluence has led to a huge demand for turf tractors for both domestic and public amenity use. This catalogue lists twelve models in three series, all but one with three cylinder diesel engines. 1998.

'Compact' is another word used to describe small tractors which can be used for domestic, amenity or light agricultural and industrial use. A full MF range of cultivation, digging and loading, and grass cutting equipment is available to fit the three 16–26hp models in this catalogue. 1983.

Three tractors of 41hp, 52hp and 60hp appeared in the 1100 series designed for use by landscapers, nurseries, building contractors, sundown farmers and municipalities. Essentially they are high horse power 'compact' tractors.

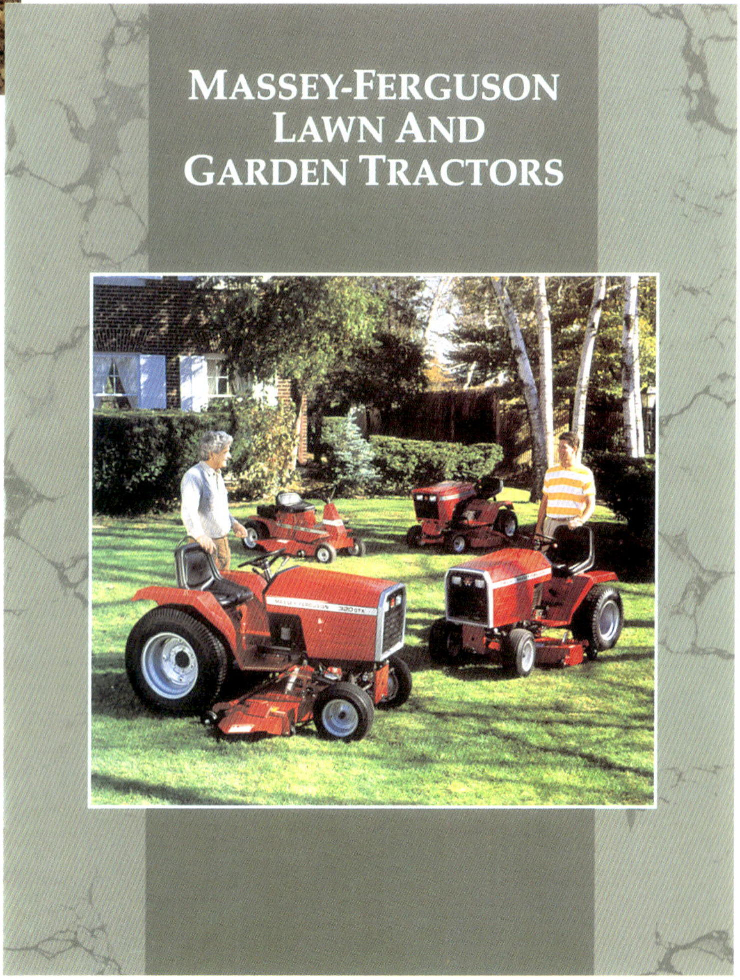

The MF 200 and 300 series of lawn and garden tractors all have hydraulic drive and were targeted at the urban lawn, suburban estates or farmsteads. There are seven models in the series of between 8hp and 20hp gasoline engines. Twenty-two implements and accessories were offered for the range. 1991.

NEW from Massey-Ferguson
Two Lawn and Garden Tractors

DICKSON COUNTY TRACTOR CO.
ROUTE 2, HIGHWAY 46A
DICKSON, TENNESSEE 37055

MF 10 and MF 12 . . . HANDIEST HELPERS AROUND

These two classic lawn and garden tractors styled on the 100 series tractors could power mowers, snow throwers, brooms, tillers, trailers, rakes, etc. 1967.

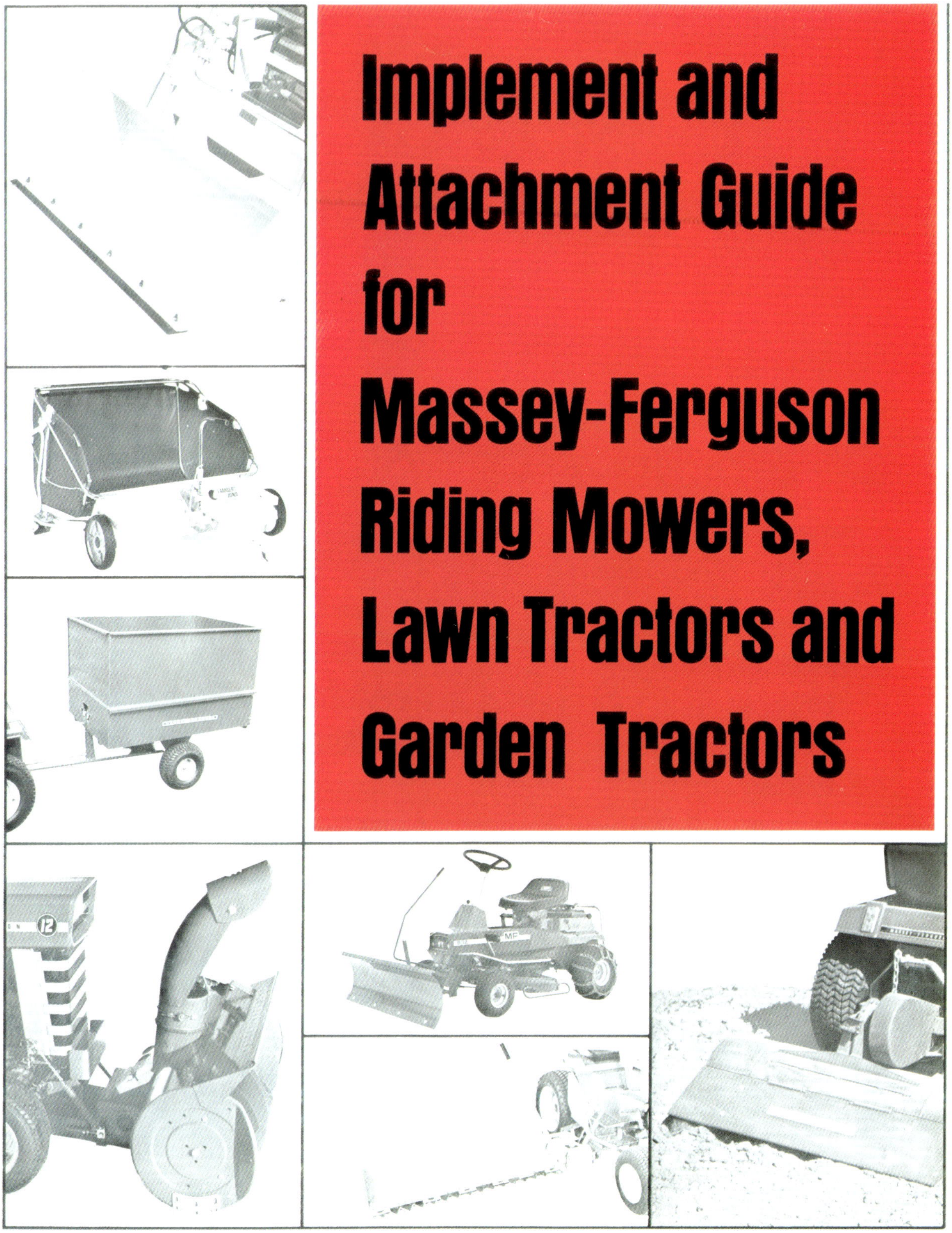

No less than 38 implements were variously offered for the MF 626/832 riding mowers and MF 8, 10/12 and 14/16 lawn and garden tractors.

MF Massey Ferguson

4-and 5-hp Rotary Tillers

MF 255 Rotary Tiller

MF 254 Rotary Tiller

Take most of the hard work out of gardening chores with a quality-built Massey-Ferguson walk-behind rotary tiller. Break up the roughest topsoil to a depth of 9". Durable, 14" bolo tines are perfect for cultivating or mulching, too.

Make quick work of Spring ground breaking in your vegetable garden with a 4-hp MF 254. Tills 26" at a swath. Vertical-shaft, Briggs & Stratton engine is a real miser on fuel. Heavy-duty transport wheels are removable for tilling at maximum depth.

If you have a larger garden, you'll want the 5-hp MF 255 – tills a full 28" path to a depth of 9". Horizontal-shaft, Briggs & Stratton engine has built-in reserve to break up stubborn, high-clay soils. Wheel height is adjustable for varying tilling/transport needs.

You'll especially like the handy Easy-Drive controls on either model tiller. Simply push a lever forward to go ahead, move to center for netural, pull back for reverse. Makes it easy to back out of fence corners or free tines of clods, rocks and debris.

Built-to-last Ultragear transmissions are geared at a 40-to-1 ratio for ease of operation and minimal engine wear. Depth control bar is easily adjusted to set desired tilling depth for different soil conditions. Four-cycle engines require *no* messy mixing of gasoline and oil – just add gasoline and go!

Massey-Ferguson rotary tillers are built to keep working, season after season. Just like reliable MF farm tractors, lawn/garden tractors and riding mowers.

Not forgetting the vegetable gardener or small professional horticulturist, MF 255 and 254 are single task machines.

Harvesting equipment

One man and harvest in the English countryside. The UK-built MF 788 combine was a development of the M-H 780. It is powered by a 56bhp Austin TVO engine or 62bhp Perkins A-4-270 diesel engine. 1961.

A small German-made machine but nevertheless tackling a relatively big field. The 31 has a 39hp Perkins A.4107 diesel engine and weighs 5200lb. Bigger models were sometimes used for harvesting experimental plots. 1961.

A German-made 685 S combine fitted with what appears to be a straw press. It has a 2.55m cut and Perkins engine rated at 62PK.

The Canadian-made Super 35 combine shown here suggests a more relaxed era of farming with time for stationary rather than on-the-move unloading! It is powered by a Continental F140-521 42hp gasoline engine.

Different models in the 400 and 500 combine series were variously made in the UK, France, Canada and Australia. The 400 and 500 were made in MF's UK Scottish factory at Kilmarnock. This machine has a Scottish registration plate. The 400 had a 72bhp engine whilst the 500 had a 94bhp engine.

M-H pioneered the use of tracks on combine harvesters, and the offering of tracks continued into the MF era. This machine was fitted with a 2m or 2.4m table and Perkins A4.235 diesel engine. It is seen harvesting rice – one of the main applications for this type of machine. 1986.

The 487 with 8hp Perkins A4.248 engine was claimed to introduce simplicity and economy to big-acre farming. It was available with fixed (2.5m, 2.9m) or telescopic (3.7m) tables and was made in France.

As with all combine advertising material, the weather is fine and apparently warm, the fields are dry and the crop is never flat or lodged! The UK-made 625 Super II with its Perkins A 6.372, 118hp engine would make light work of the crop here. 12ft, 14ft and 16ft tables were fitted.

Since the beginning of farm mechanization, manufacturers have been under constant pressure to improve performance and come up with innovative modifications and developments. The Power-Flow table for combine harvesters was one such example. It was said to solve the problem of obtaining an optimum distance between the cutter bar and auger. The Power-Flow table was claimed to improve output by 73 per cent in direct cut oilseed rape, 12 per cent in barley and 15 per cent in wheat.

This 825 combine is seen working in a heavy wheat crop in continental Europe. It has a six cylinder, turbo charged Perkins AT6.3544 133hp engine and a quick attach table which was available in 3.05m, 3.66m and 4.27m widths. 1983.

The last two decades have seen most combine manufacturers developing rotary type combine harvesters. MF has played its part in this. This 240hp Cummins engined model can be fitted with up to 30ft wide headers. This North American shot shows well its capacity and suitability for large acreages. 1993.

Adapting new technology to agriculture. MF pioneered the electronic monitoring of cereal yield as it is being harvested, and the translation of this into field yield maps. In this way, farmers can identify where yield variation occurs within fields. The system involves one of the satellite Global Positional Systems. 1995.

Progressive development of electronic cereal yield mapping led to 'Fieldstar'. Once reasons for yield variation within a field have been established, the Data Vision equipment can be transferred from combine to tractor. In this way if, for example, available soil nutrient variability has caused yield variability, then the Data Vision equipment can be used to variably control fertilizer application to selected areas within a field. The system has been awarded the Royal Agricultural Society of England's gold medal. 1996.

Natural pasture and cultivated forage crops for livestock are the world's largest crop in terms of area. Reciprocating blade mowers, first made in the early 19th century, were basically the only form of mower for grass and cereals for well over a century, and are still made today. This 1960s' UK 5ft mower is working in a heavy crop destined for hay.

New concept, high efficiency rotary mower.

MF70

Late 20th century heavy use of nitrogen fertilizers produced crops that are too dense for reciprocating blade mowers. A new generation of revolving disc or drum mowers came into being. They have high output rates and rarely block. The drums on this mower rotate at 1766rpm.

Both MH and Ferguson marketed forage harvesters and they have become an important MF line of equipment with associated trailers. This machine needs 75–120hp, has a chop length which can be varied between 10mm and 40mm, and an output of up to 60t/ha of wet grass given a good trailer team! 1989.

Just as combine harvesters developed from trailed to self-propelled, so have forage harvesters. They are large machines, more often than not operated by contractors. The largest machine in this series has a 480hp engine and weighs over 9t. 1996.

MF 220 Swather

88 hp, triple delivery

MASSEY FERGUSON

Another machine which has evolved from trailed to self propelled is the swather. Used for cutting grain or forage crops, and leaving them in a swath to dry prior to being picked up by a combine, baler or forage harvester, they have high outputs. This 88hp machine can be fitted with a range of headers up to 30ft wide. 1996.

M-H and Ferguson both sold a simple basic model of a baler producing small rectangular bales. The first pair of MF balers were the 56in. pick up MF 10 and 48in. pick up MF 703.

Small rectangular type balers held the market for many years with progressively increasing outputs. However, demand for efficiency and total mechanical handling led to the development of large round balers which can cope with hay or silage crops. The largest machine in this series requires 50hp and produces a 1.3–1.9m x 1.2m bale of up to 748kg. 1995.

Demand for ever larger bales has led to the development of large rectangular balers. The 190 model needs 135 pto hp to produce bales of 1.18m x 1.27–2.8m. A straw bale can weigh up to 600kg. 1996.

Back to a simple black and white photo for one of the latest introductions to MF's baler range in the 130 'centre line' series. Here the advantages are clearly claimed. In the 1960s MF in France made a 'straight through' low density baler, but it was towed in a conventional offset mode. 1995.

In-field wafering or cubing of hay crops, particularly alfalfa as in this shot, has received attention from time to time from various manufacturers. MF made this one but apparently few were sold and the line was discontinued – perhaps surprisingly given the 'dazzling' end product!

C. L. MILNER

THE REVOLUTIONARY

MASSEY-FERGUSON

515 CANE HARVESTER

THE ONLY MACHINE IN THE WORLD THAT CUTS, TOPS AND LOADS CANE IN ONE OPERATION

MF

the Massey-Ferguson system of mechanised farming

Mechanical cane harvesting was very much an Australian MF operation. To gain the expertise necessary to cut through cane 'forests' like this, they purchased the Crichton Co. in 1966. First models were tractor mounted, with self-propelled versions made later.

Massey-Ferguson Potato Harvester 711

Single row machine with low power requirement. Bulk delivery and bagger models

Both M-H and Ferguson marketed spinner diggers for potatoes, and M-H also sold elevator diggers. All required a team of hand pickers. As far as can be ascertained, MF only made one venture into the potato harvester market with this 711 machine, shown working in rare, near perfect lifting conditions in the UK.

1 Whole Crop Harvester

MASSEY FERGUSON

A new concept in harvesting

Finely broken high density straw — ideal for animal feed, is collected in bags

Can be used for stationary threshing — versatility which provides maximum return on investment.

Clean grain is collected in bags.

Fitted with a 2 metre table and operated in reverse allows a field to be opened without the tractor wheels running down the crop.

Can be transported within the width of the tractor, ideal on narrow lanes, etc.

This unusual machine was developed for those countries where wheat is a major crop and where finely broken straw is required for animal feed. The broken straw is collected in bags or blown into a trailer. 1991.

Industrial equipment

The UK MF 35 tractor carried on the industrial design traditions established in the TE Ferguson era. The one illustrated here is a de luxe four cylinder diesel engine type with obviously a very relaxed and contented operator.

Both Ferguson and M-H had produced industrial tractors – basically modified agricultural tractors. Industrial tractors were always deemed to be suitable for 'tougher' conditions and indeed often had rigid or stronger front axles. Here we see the MF 65 tractor adopting the yellow livery which was to characterize the later extensive range of MF equipment.

The MF 702 Tractor-Digger-Shovel broke new ground when it was launched in 1959. It is based on the Ferguson 35 tractor with a four cylinder Standard 23C, 37.4bhp diesel engine. It is equipped with the 11 cu ft 702 loader and 710 backhoe with a digging depth of 12ft.

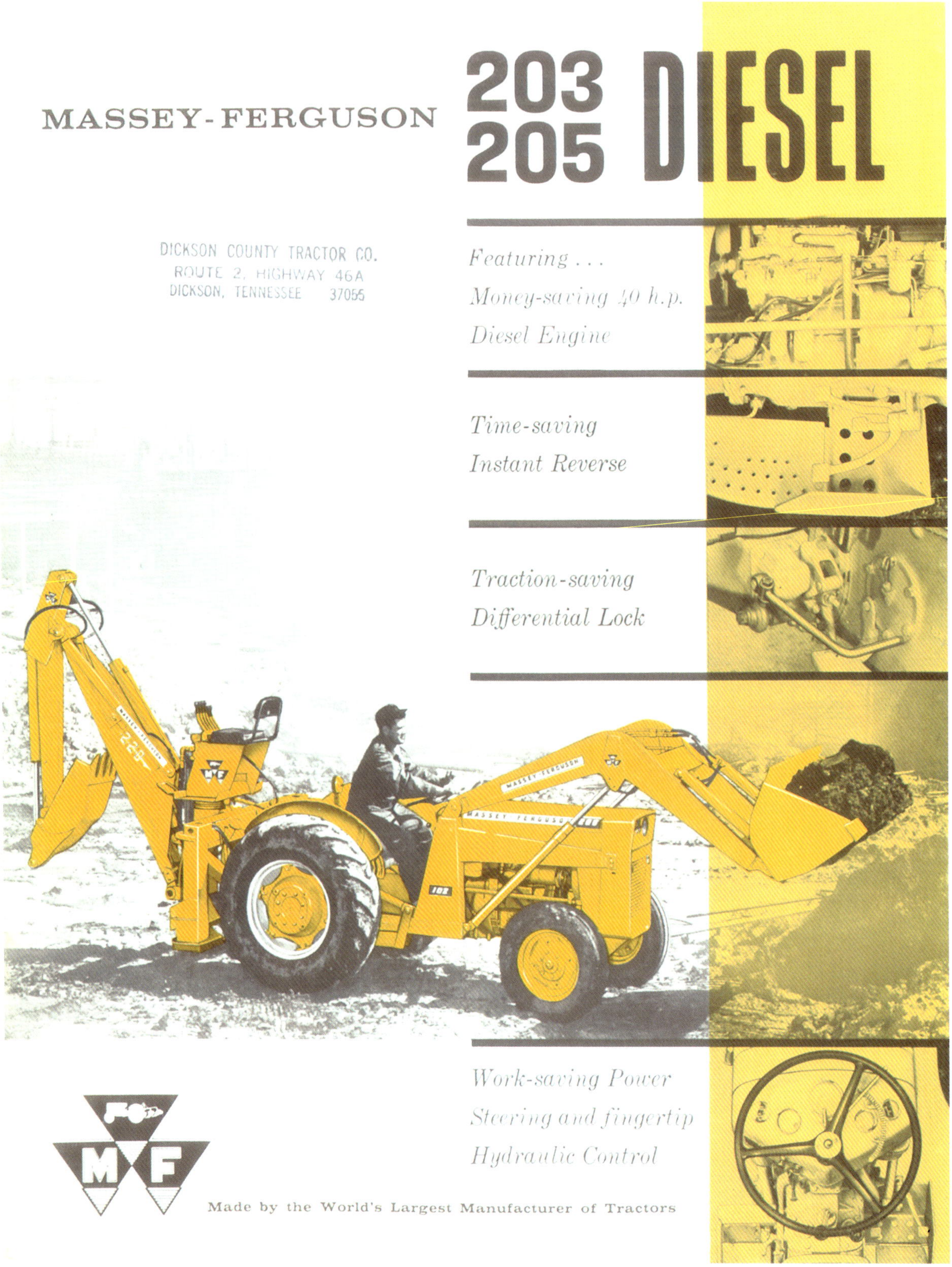

The 203 and 205 tractors, shown here in North American advertising were both powered by Perkins 3.A.152 three cylinder diesel engines. The 203 had a standard six forward, two reverse gearbox, whilst the 205 had a torque converter and hydraulic shuttle transmission with four forward and four reverse gears.

The 100 and 200 loaders were of classic Davis design – the company MF had originally purchased to get into the industrial equipment business. The lighter 100 loader was commonly used on American MF 35 tractors, and the heavier 200 on 203, 205 or 65 Utility tractors. 1962.

The establishment by MF of an Industrial and Construction Machinery Division in 1966 heralded MF's intention to expand dynamically into the industrial machinery market. This reached its peak in the 1970s with the opening of new factories for this purpose in North America and Europe. For the USA market a new series of detailed *Buyer's Guides*, solely for industrial equipment, was introduced. 1972.

The 1979 *Industrial Machinery Buyer's Guide* shows a more modern, perhaps less interesting style than that for 1972.

Massey-Ferguson MF222/MF232 Backhoes

Center-mount or hydraslide/ 13', 14', 15' digging depth combinations

Backhoes came in two principal forms – centre-mount or the perhaps more familiar side-shift type. Centre-mounts are more commonly used when there is no requirement to work close up against walls etc. The 222/232 backhoes were made for MF 235, 200, 3165 and 300 tractors.

MF
Massey Ferguson

MF 200B Crawler Loader with Digger

MF made a wide range of crawler shovels and crawler loaders. This 200B is unusually fitted with an MF 220 backhoe. The machine is powered by a Perkins A3.152 three cylinder diesel engine.

Although early industrial machines were simply adaptations of farm tractors, farms lagged behind the construction business in adopting heavy duty industrial and construction equipment. However, increasing farm size and the costs of labour have led to farmers adopting materials moving equipment modified from industrial equipment. This telescopic handler offers 72hp. 1984.

Another adaptation of industrial equipment to the farm is this 80bhp four wheel drive Farm Handler which weighs 4.5t. 1989.

The *MF News* contained occasional supplements on MF Industrial equipment which carried features on new products, and MF industrials working in very varied situations around the world. This 350 C tracked excavator was also available as an all wheel drive excavator. 1978.

MF
Massey Ferguson

MF 450 S EXCAVATOR

- 9'8'' (5994mm) digging depth
- 27,500 pound (12473 kg) bucket force permits digging in extremely hard materials
- Transverse engine affords good visibility and is part of the counterweighting
- Sliding hood and hinge panels permit easy and quick engine service
- Crawler tractor type undercarriage provides longer track life
- Full hydrostatic transmission with separate radial piston motors for each track permits counter-rotation
- Six-position adjustable boom allows unit to be matched to the job
- Low profile design makes it easy to haul the unit
- Smooth operating, six cylinder engine provides 93 horsepower

The 450 S tracked excavator has a Perkins A6-354, 93 flywheel hp engine and has a maximum digging depth of 19ft 6.25in. 1974.

MF

Massey Ferguson

MF Compactors

For waste control

The 27 ton, 206hp, C666C (top) was the biggest of MF's range of three compactors. This model was also produced as MFís only wheeled bulldozer – the D66C.

The *MF Product Guide* series was produced for European countries. This brochure lists specifications of four industrial tractors, five tractor loaders, eight backhoe loaders, two fork lifts, eight wheel loaders, three refuse compactors, seven crawler loaders, seven crawler dozers, two tracked excavators and two wheeled excavators. The 210hp 66C wheel loader shown here weighs 18,660kg and has a breakout force of 12,300kg.

The MF 11 front wheel drive loader was the last of just a few of these types of MF machines. This once popular machine is 74hp and has an operating capacity of 4000lb with a breakout force of 10,000lb. 1975.

Customer demands for industrial equipment to work in difficult terrain like this caused MF to adopt four wheel drive on many of its industrial machines. MF ultimately produced over 50 models of tractor backhoe loaders, of which the 50B shown here was one of the most successful models.

Massey Ferguson

MF20F

Turf/Utility

MF Industrial

These Turf/Utility tractors were produced in industrial colours but were meant for light duties. Low ground pressure is a feature of such machines. The 20F (top) is seen fitted with 21.5 x 16.1 rear and 12/12 x 26 fronts – big tyres for a tractor weighing only 4166lb. 1986.

In the last decade the demand for small, compact but rugged industrial equipment has soared with ever increasing labour costs. This 2.8t, 22.5bhp mini excavator can be fitted with a wide range of equipment in order to work in very confined spaces.

In 1967 MF's industrial division made a policy decision to move into logging equipment. The Treever was one such product. Powered by a Perkins A3.152 engine, the loader has a lift capacity of 1500lb and the carrier a two cord capacity.

In 1983 MF industrial equipment manufacture became focused on the Manchester, UK factory, after most other industrial plants had closed. The operation was re-named MF Industrials. In 1992, in a management buyout, this was sold to FERMEC who continued to use the MF badge. In 1996 FERMEC was bought by the Case Corporation and MF badging of equipment ceased in early 1997. This 865 tractor backhoe loader was one of the last MF badged machines and sports the air conditioned cab comfort of the late 20th century. This is believed to be the last 'MF Industrial' brochure from FERMEC.

Cultivation and seeding equipment

It would require several books to list all the ploughs which have been made in the history of MF. There have been perhaps many more models of the plough than any other implement. This large plough demonstrates how, with increasing tractor power and plough size, a compromise of 'semi-mounted' over the original Ferguson concept of fully mounted has had to be made due to plough weight and length. As always we see a plough working in ideal conditions and beautiful soil!

The only alternative to mouldboard ploughs for achieving a deep cultivation and significant soil inversion is disc ploughs. They have better penetration in hard dry conditions, but are less good at burying trash. This brochure is bilingual for use in Arabic-speaking areas. 1984

MF
Massey Ferguson

MF27 SUBSOILER

Improve drainage... aerate for high yields

The method of deepest cultivation is subsoiling. Whilst Ferguson marketed one with a C blade, M-H apparently never made a subsoiler. This early MF reversible blade version was made by MF Australia and penetrates up to 19in. It can be fitted with a mole drainer or pipe laying attachment. The absence of wheelmarks here belies the power and traction required to pull a subsoiler!

A UK-made 88hp four wheel drive articulated tractor pulling a three tine subsoiler through cereal stubble. It is operating in good dry conditions which are optimal for the operation. This implement can penetrate to a depth of 23in.

the Massey-Ferguson system of mechanised farming

Perhaps the second most numerous types of equipment after ploughs are tine cultivators and toolbars. Who knows, they may even outnumber ploughs. The Australian-made MF machine with coil-spring tines shown here is being used in a young sugar cane crop.

One of MF's oldest 'licensed' producers of equipment is TAFE in India – a relationship that goes back to the assembly of UK-made Ferguson tractors in India, and early ties with Perkins engines. The relationship is still very strong with MF derivative tractors and implements and Perkins engines being produced to the present day – including MF 35 style tractors.

High quality, high output tandem disc harrow

MF520

Disc cultivators first entered the Massey stable when M-H purchased the Corbin Disc Harrows business of Prescott, Ontario in 1893. The most common type has been tandem gangs which have progressively increased in size with tractor power. This trailed example is being pulled by an MF 100 series tractor equipped with a pressure controlled multipull type of hitch.

MASSEY-FERGUSON TRASHEATER 740

- 4" x 4" Tubular Frame
- 15" Shank Spacing
- 26" x 32" Shanks
- 1" or 1¼" Thick Shanks
- Spring or Rigid-Mounted Shanks
- ⅝" x 2" x 16" Reversible Chisels
- 15" x 6" Rims (Use 7.60 x 15" Tire)
- 20" Coulter Blades
- 9" Coulter Spacing
- 1⅛" Square Axles
- Flexible Bearing Hangers
- Triple-Sealed Regreaseable Ball Bearings
- Easy Adjusting Coulter Gangs
- Trash Bars

OPTIONAL

- One-Piece Helical Blade
- Three-Piece Helical Blade
- Twisted Shovel
- 15" x 10" Flotation Rims
- 7.60 x 15" Tire and Tube
- 11L x 15" Tire and Tube
- Hydraulic Hose and Cylinder Kit
- 1⅛" Gang Bolt Wrench
- Weight Holders
- 15" x 8" Rim
- 9.5L x 15" Tire and Tube

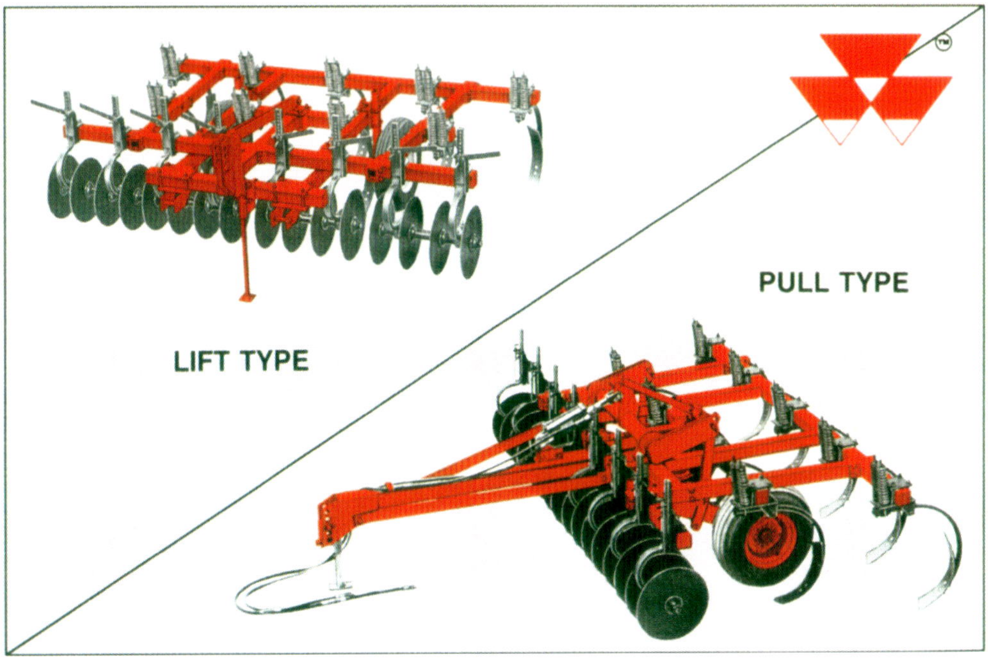

Non-plough cultivation systems require cultivation equipment which can incorporate previous crop residues into the soil surface. One such machine is the Trasheater which combines the use of discs and tines. Here we get hard facts in an unemotional presentation of the specifications.

Almost 1400 dollars per furrow for this plough. Obviously the cost was justified on such rich deep soils!

An M-H USA-built tractor pulls an MF Australian-built drill cultivator, with extra cover for the seeds coming from the diamond back harrows trailed behind. Note the continuing use of the Sunshine brand name even in 1957.

An early type of direct drill made in Australia.

Certain combinations of soil and climatic conditions permit direct seeding of small grain cereals into the previous crop stubble. This 130 drill weighs about 2t and is 11ft 10in. wide. MF designed the first such drills in UK 15 years before this model appeared.

Many attachments could be fitted to this South African-built MF toolbar. Here it is seen made up as a two row ridge planter/fertilizer unit. The listers produce flat top ridges and fertilizer is placed behind coil shank openers. Lister beams and bottoms are particularly suited to heavy cotton soils. 1962.

A modern implement of conventional design. This combined planter and fertilizer unit typifies a type of planter widely used in North American maize, soya, sunflower and cotton areas. This model was made in 2-6 row versions. 1989.

The use of the pneumatic principle to move seed from box to coulter has led to the design of seed drills with a wide seeding width but narrow seed box. This machine has a 19ft 8in. sowing width and carries almost 1.2t of grain. 1988.

Other equipment

Having influenced the production of grain for over a century, MF decided to enter the grain storage and handling equipment market in the 1960s, but the venture has not survived to the present day. The equipment shown here was marketed by Massey-Ferguson (Farm Services) Ltd. with a little emphasis on MF standard red and grey livery! 1965.

BUTLER PANL-FRAME FARM BUILDINGS

In the 1960s MF also entered the farm building sector by linking with Butler to form Massey-Ferguson-Butler Ltd., which was based in Coventry, UK. Panl-Frame buildings were recommended for pig units, beef production and machinery storage and could be up to 40ft wide in clear span form. 1963.

Ferguson originally developed a small two wheel, hydraulic tipping trailer for the Ford Ferguson tractor. The two wheel principle placed weight on the tractor for added traction. This is a 1994 'up-grade' of Ferguson's original concept.

Tractor loaders became commonly available for latter-day M-H tractors with external hydraulic pumps, and for post-war Ferguson tractors using the internal hydraulic pump. Through the years loader capacity has increased with tractor size and power. This 1983 MF loader has a maximum load capacity of 1200–1310kg.

As far as can be established, the four models of MF power washers were only marketed in North America – together with a range of MF branded cleaning chemicals, concrete floor cleaners, sanitizers, spray waxes and degreasers.

At least eight chain saw models were branded by MF including one with electric drive. They appear to have been marketed in the late 1970s.

M-H began selling horse-drawn sledges for domestic, farm and forestry use in the late 19th century. Personal travel on snow was later mechanized with this four model Ski Whiz snowmobile range. Other MF snowmobile ranges were the Cyclones, Whirlwind, Chinook and Storm. 1971.

MF Massey Ferguson

8-hp Snow Blower

MF 228 Snow Blower

Turn back-breaking, heart-pounding snow removal into an easy walk with a self-propelled Massey-Ferguson Snow Blower. The MF 228 clears a 26″ wide path through waist-deep snow. Two-stage operation assures efficient removal of the most stubborn drifts. High-capacity auger chops hard-packed snow into smaller pieces, feeding it into a high-speed impeller. Snow is thrown out of the directional discharge chute up to thirty feet away.

Adjust direction of throw from the operator position of the MF 228 with a handy chute control crank. Swivels chute from 90° right to 90° left. Handle-mounted lever offers a choice of four speeds forward, neutral and reverse.

The 8-hp Tecumseh engine is fully winterized with an oversized recoil starter for easy starting, hooded gas cap, carburetor heater box, moisture-shielded spark plug and float-type carburetor with primer.

The MF 228 is built to tackle the toughest snow-clearing jobs. And designed for ease of operation with minimum maintenance. Easily replaceable shear bolts protect gears from damage if you strike an unexposed object. Adjustable handle height. Auger sides are fully enclosed for greater efficiency and safety. Four-position, free-wheeling depth gauges help assure uniform operation.

For domestic or public pathway clearance, this 8hp snow blower enabled one to 'turn back-breaking heart-pounding snow removal into easy walk'. The 228 can clear a 26in. path through 3ft drifts and throw the snow 30ft away. 1974.

Massey Ferguson

SNOWBLOWERS

Blow snow in a big way from the comfort of a tractor cab. Six models were available in this range, ranging from 12–30hp up to 80–125hp. Cutting width of the largest model was 108in. and cutting height 42in. with a double auger – as shown here.

930 Powerstations

22kVA to 880kVA

MASSEY FERGUSON

MF
Massey Ferguson

Forty-four models of power stations were offered in this 1994 brochure ranging from 22 kVA to 880 kVA. All are powered by Perkins engines. Let there be light!

General publicity and celebration material

Advertising by provision of information was the theme of this 1970s guide. It gave guides to implement type and tractor power suitability for irrigated and dryland farming conditions, together with husbandry guidelines for some principal crops. Great results if you follow the guidelines

Ambassadors of service

Good after-sales service for equipment sold is an old strategy employed to nurture customer satisfaction and brand loyalty. In 1998 AGCO, the new owners of MF, were still hard at the task – with a trusty Land Rover – pressing home the message to their dealers in *Voice* – an MF magazine for MF sales staff, distributors and customers.

HERITAGE

This 50-page celebration brochure was issued in 1972 for the 125th anniversary of the company. It is a must for any Massey enthusiast, with a surprising number of copies still turning up on the second-hand literature market.

The following five period reproduction posters were issued for dealer distribution to farmers in the UK in order to celebrate 125 years of the company. The original size of the posters was 18 x 24in.

MASSEY-HARRIS No. 3 STEEL RAKE
WIDTHS
24 TEETH 8FT
30 TEETH 10FT 36 TEETH 12FT

MASSEY-HARRIS No. 7 MOWER
WIDTHS OF CUT

MASSEY-HARRIS CULTIVATOR

MASSEY-HARRIS
MASSEY-HARRIS
138, Rue d'Allemagne, PARIS

THE MASSEY-TORONTO BINDER
OBTAINS THE
HIGHEST HONORS
AT THE GREAT
INTERNATIONAL
FIELD TRIAL
NOISIEL.
JULY, 19 TO 22, 1889
WORLDS
EXPOSITION
PARIS-FRANCE
ARCH OF TRIUMPH
GRAND OBJECT OF ART & GOLD MEDAL
TORONTO
LIGHT
BINDER
TORONTO
BEING A
GRAND
OBJECT
OF
ART
AS USUAL PROVED TO BE
LIGHTEST IN DRAFT, SUREST TO TIE,
EASIEST TO MANAGE,
AND THE FIRST TO FINISH ITS PLOT IN EVERY TEST.
ALSO
GOLD MEDAL
WAS AWARDED
THE
MASSEY
EXHIBIT
MASSEY M'F'G. CO.
N.B.—This Company also obtained the highest award, a GOLD MEDAL,
and 400 francs, at the great International Field Trials at Saintes, France,
held on July 20th, 1890.

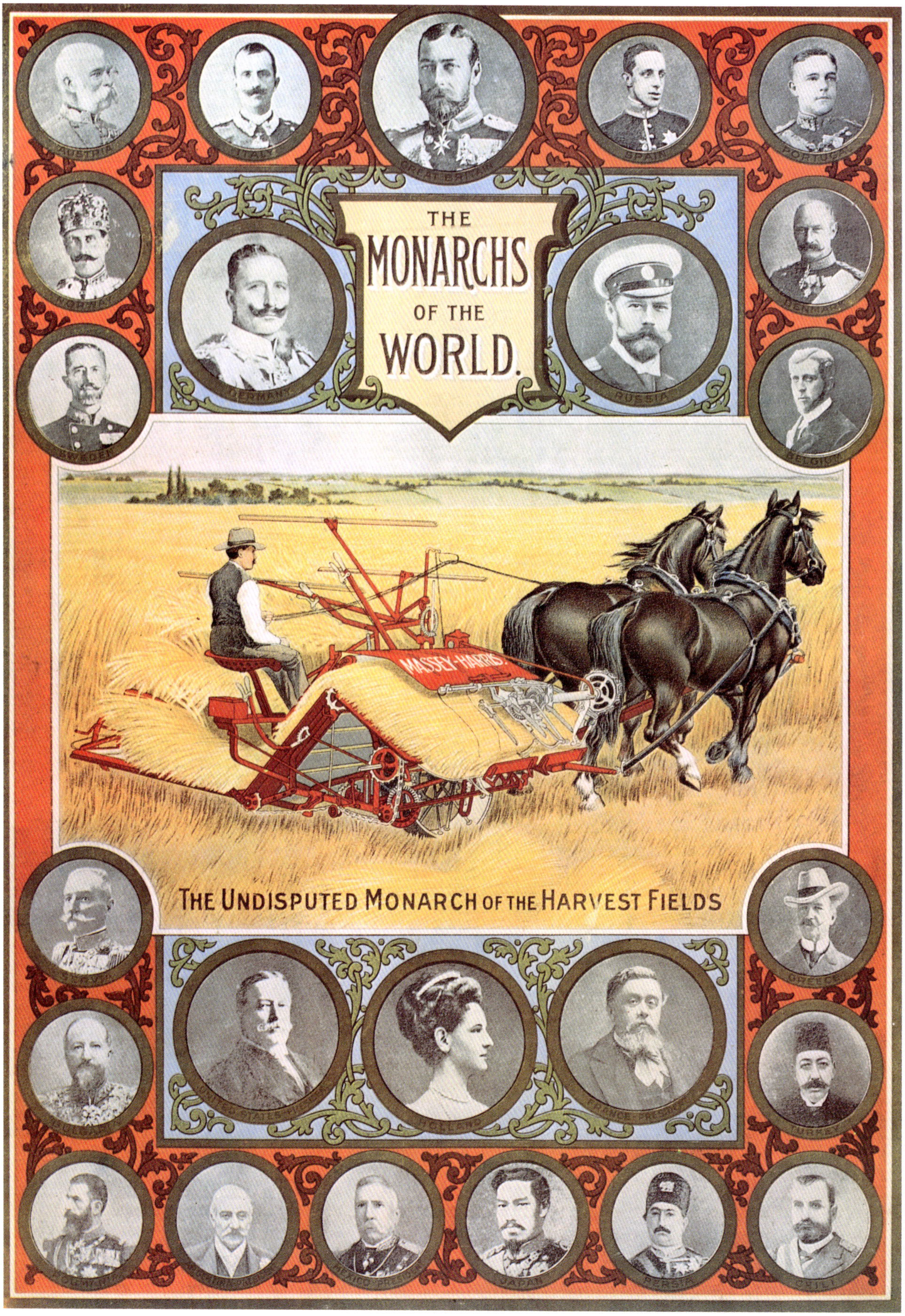
THE
MONARCHS
OF THE
WORLD.
GERMANY
RUSSIA
SPAIN
MASSEY-HARRIS
THE UNDISPUTED MONARCH OF THE HARVEST FIELDS
GREECE
HOLLAND
JAPAN
PERSIA

Tastefully and technically celebrating the 50th anniversary of the start of Ferguson tractor production in the UK in 1946 on the front of the MF *Voice* magazine. Harry Ferguson looks on …

Production milestones have traditionally been causes for celebration – but nothing in MF history has surpassed the half million milestone achieved by the UK-built Ferguson TE tractor.

150 Years of Massey Ferguson Heritage

Harris Brantford Binder (1890)

150
1847 – 1997

MASSEY FERGUSON®

Serving World Agriculture
Since 1847

MF 8160 Tractor (1997)

THE COMPANY
HERITAGE

MASSEY FERGUSON

North American and UK celebration leaflets of 150 years of the company. 1997

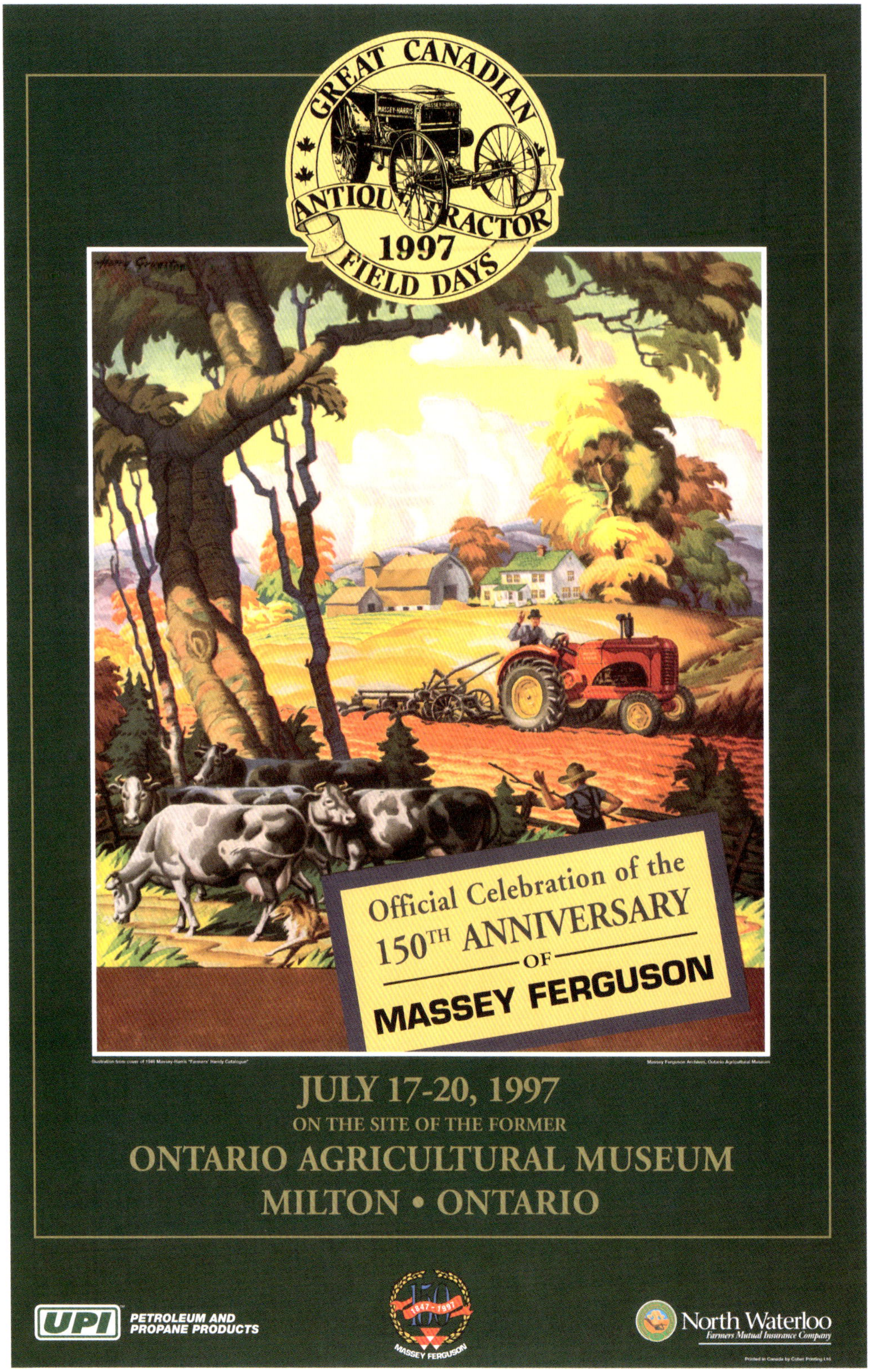

Ending where we began! Beautiful artwork reproduced for the company's 150th anniversary (from the cover of the 1946 *Farmers Handy Catalogue*). Men, machines and livestock working together in a bygone era are shown on this poster (original size: 22 x 33in.) which is already a collector's item. Please note that for convenience the surrounding plain border has been removed.

INDEX